# Thinking about Acting

# Thinking about Acting

## *Logical Foundations for Rational Decision Making*

John L. Pollock

OXFORD
UNIVERSITY PRESS

2006

OXFORD
UNIVERSITY PRESS

Oxford University Press, Inc., publishes works that further
Oxford University's objective of excellence
in research, scholarship, and education.

Oxford New York
Auckland Cape Town Dar es Salaam Hong Kong Karachi
Kuala Lumpur Madrid Melbourne Mexico City Nairobi
New Delhi Shanghai Taipei Toronto

With offices in
Argentina Austria Brazil Chile Czech Republic France Greece
Guatemala Hungary Italy Japan Poland Portugal Singapore
South Korea Switzerland Thailand Turkey Ukraine Vietnam



Published by Oxford University Press, Inc.
198 Madison Avenue, New York, New York 10016

www.oup.com

Oxford is a registered trademark of Oxford University Press



Library of Congress Cataloging-in-Publication Data
Pollock, John L.
Thinking about acting : logical foundations for rational
decision making / John L. Pollock.
p. cm.
Includes bibliographical references and index.
ISBN-13 978-0-19-530481-7
ISBN 0-19-530481-0
1. Statistical decision. 2. Decidability (Mathematical logic)
3. Probabilities. 4. Induction (Logic) I. Title.
QA279.4.P65 2006
519.5'42—dc22 2005054921

1 3 5 7 9 8 6 4 2

Printed in the United States of America
on acid-free paper

For Lilian

# Preface

The objective of this book is to produce a theory of rational decision making for realistically resource-bounded agents. My interest is not in "What should I do if I were an ideal agent?" but rather, "What should I do given that I am who I am, with all my actual cognitive limitations?"

The book has three parts. Part I addresses the source of the values that agents use in rational decision making. The most common view among philosophers and cognitive scientists is that the primitive evaluative database that real agents employ in evaluating outcomes is a preference ranking, but I argue that this is computationally impossible. An agent's evaluative database must instead assign real numbers to outcomes. I argue that, contrary to initial appearances, this is psychologically plausible.

Part II investigates the knowledge of probability that is required for decision-theoretic reasoning. I argue that subjective probability makes no sense as applied to real (resource bounded) agents. Rational decision making must instead be based on a species of objective probability. Part II goes on to sketch a theory of objective probability. I use that to define a variety of causal probability and argue that this is the kind of probability presupposed by rational decision making.

Part III explores how these values and probabilities are to be used in decision making. Classical decision theory is based on the optimality principle, according to which rationality dictates choosing actions that constitute optimal solutions to practical problems. Optimality is defined in terms of expected values. I will argue that the optimality prescription is wrong, for several reasons: (a) actions cannot be chosen in isolation—they must be chosen as parts of plans; (b) we cannot expect real agents to find optimal plans, because there are infinitely many alternatives to survey; (c) plans cannot be evaluated in terms of their expected values anyway, because different plans can be of different scopes. I construct an alternative, called "locally global planning", that accommodates these difficulties. According to locally global planning, individual plans are to be assessed in terms of their contribution to the cognizer's "master plan". Again, the objective cannot be to find master plans with maximal expected values, because there may be none, and even if there are any, finding them is not a computationally feasible task for real agents. Instead, the objective must be to find good master plans, and improve them as better ones come along. It is argued that there are computationally feasible ways of doing this, based on defeasible reasoning about values and probabilities.

This work is part of the OSCAR project, whose objective is to construct an implementable theory of rational cognition and implement it in an AI system. This book stops short of implementation, but that is the next step. This book provides the theoretical foundations for an implemented system of decision-theoretic planning, and future research will push the work in that direction.

Much of the material presented in this book has been published, in preliminary form, in other places, and I thank the publishers of that material for allowing it to be reprinted here. Much of Part I is drawn from "Evaluative cognition" (*Nous*, **35**, 325–364). Chapter 8 is based upon "Causal probability" (*Synthese* **132**, 143–185). Chapter 9 is based upon "Rational choice and action omnipotence" (*Philosophical Review* **111**, 1–23). Chapter 10 and part of chapter 12 are based upon "Plans and decisions" (*Theory and Decision* **57**, 79–107) and "Against optimality: Logical foundations for decision-theoretic planning in autonomous agents" (*Computational Intelligence* **22**). The appendix is a revised version of "The theory of nomic probability" (*Synthese* **90**, 263–300).

I also thank the University of Arizona for its support of my research. I particularly want to thank Merrill Garrett for his continued enthusiasm for my work and the help he provided in his role as Director of Cognitive Science, and I want to thank Chris Maloney for his steadfast support as Head of the Department of Philosophy. I am indebted to numerous graduate students for their unstinting constructive criticism, and to my colleagues for their interactions over the years. I want to mention specifically Douglas Campbell, Josh Cowley, Justin Fisher, and Nicole Hassoun, who helped me more than I can say.

This work has been supported by grants no. IRI-9634106 and IRI-0080888 from the National Science Foundation.

# Contents

# Part III: Decisions

# Thinking about Acting

# 1
# Rational Choice and Classical Decision Theory

## 1. Rational Cognition

We make decisions constantly, at almost every moment of our waking lives. Most are little decisions—"Should I put more mustard on my sandwich?" But some are momentous—"Should I marry Jane?" Some people are better decision makers than others, and some decisions are better than others. What makes one decision better than another? One sense in which a decision can be better is that it has a better outcome. But there is also an internal dimension of criticism. A decision can be evaluated as having been made well or badly regardless of its outcome. Because Claudio was furious with Flavia, he spent his paycheck on lottery tickets rather than paying the mortgage. He got lucky and won, and they are now millionaires, but it was still a stupid thing to do. His decision was irrational.

What makes a decision rational or irrational? How should we go about making decisions so that they are rational? That is the topic of this book. I want to know how we, as human beings, should go about deciding what actions to perform.

We are cognitive agents. Cognitive agents think about the world, evaluate various aspects of it, reflect upon how they might make it more to their liking, and act accordingly. Then the cycle repeats. This is the *doxast-conative* loop, diagrammed in figure 1.1. The defining characteristic of cognitive agents is that they implement the doxastic-conative loop by thinking about the world and acting upon it in response to their deliberations. Both human beings and the autonomous rational agents envisaged in AI are cognitive agents in this sense.

This cognition can be divided roughly into two parts. *Epistemic cognition* is that kind of cognition responsible for producing and maintaining beliefs. *Practical cognition* evaluates the world, adopts plans, and initiates action. We can further divide practical cognition into three parts: (1) the evaluation of the world as represented by the agent's beliefs, (2) the selection of actions or plans aimed at changing it, and (3) the execution of the plans.

Some aspects of our cognition are beyond our control, and it makes no sense to ask how we should perform those cognitive tasks. For example, when I look at the world, purely automatic computational processes take as

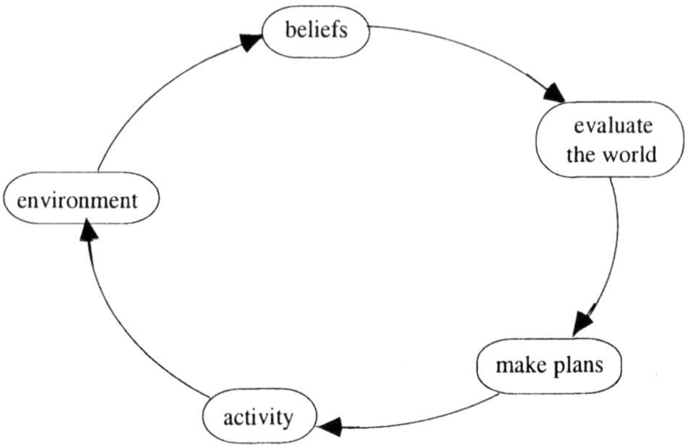


Figure 1.1 The doxastic-conative loop

input the pattern of stimulation at my optic nerve and produce a visual image. The visual image provides my visual access to the world. But I have no control over how the image is produced. If I see a newspaper illuminated by what I know to be red light, and the newspaper looks red to me, I cannot be criticized as irrational because it looks red to me. That is beyond my control. But I can be criticized as irrational if I believe on the basis of the visual image that the newspaper really is red. This is because the inference is something over which I have a certain amount of control. I can at the very least withdraw my conclusion in light of my knowledge that newspapers are generally white and my knowledge of how red lights can make white things look red. But there is nothing I can do to make the newspaper stop looking red to me.

A theory of rationality is a theory about how a cognitive agent should perform the kinds of cognitive tasks over which it has some control.[1] Just as cognition divides roughly into epistemic cognition and practical cognition, so rationality divides roughly into epistemic rationality and practical rationality. Epistemology studies epistemic rationality, and I have written about that extensively elsewhere.[2] The focus of this book is practical rationality. I want to know how a cognitive agent should go about deciding what actions to perform. An answer to this question constitutes a *theory of rational choice.* So this is a book about rational choice.

My principal concern is with human decision making. I want to know how we, as human beings, should decide what actions to perform. However, idiosyncratic features of human psychology sometimes obscure the logic of

[1] This point is developed more fully in Pollock (2006).

[2] See particularly my (1986, 1995) and Pollock and Cruz (1999).

rational decision making, and we can often clarify the issues by focusing more broadly on rational decision making in any cognitive agent, human or otherwise. Humans are the most sophisticated cognizers we currently know about, but we can usefully ask how *any* cognitive agent should go about deciding how to act. The results of this investigation should be as applicable to the construction of artificial rational agents in AI as to human beings. The advantage of taking this broader perspective is that it can sometimes be argued that purely computational considerations illuminate issues in the theory of rational choice, showing that theories motivated by thinking specifically about human beings cannot be correct for any cognitive agents, and so in particular they cannot be correct for human beings.

The term "practical reasoning" has been used ambiguously in philosophy, on the one hand to refer to purely self-interested reasoning about action, and on the other hand to include the moral aspects of decision making. As I use the terms "practical reasoning" and "practical cognition" in this book, they are about purely self-interested decision making. An individual comes to a decision problem with various goals and then tries to select actions that will achieve those goals. I want to know how such decisions should be made. The "should" here is a practical "should", not a moral "should". The problems of morality are orthogonal to understanding practical cognition in this sense. Morality could interact with practical cognition in various ways. It *might* function by simply adding goals to be achieved by practical cognition, or by affecting the evaluation of goals. In either case, morality would function via the mechanisms of practical cognition, and would not be in conflict with it. But morality might also function in a way that puts it in potential conflict with self-interested practical decision making. Moral philosophers have endorsed both of these views of the relationship between morality and practical decision making. In this book, however, I propose to remain neutral on issues of morality.

## 2. Ideal Rationality and Real Rationality

Human beings, and any real cognitive agents, are subject to cognitive resource constraints. They have limited reasoning power, in the form of limited computational capacity and limited computational speed. This makes it impossible, for example, for them to survey all of the logical consequences of their beliefs, or to compare infinitely many alternatives. This is a fundamental computational fact about real agents in the real world, and I would suppose that it could not have been otherwise. An account of how a real agent should make decisions must take account of these limitations.

Theories of rational action are sometimes taken to be theories about how ideal agents, immune to such cognitive limitations, should make decisions (Cherniak 1986; Skyrms 1980, 1984; Lewis 1981). One can, of course, choose to talk that way, but it is hard to see what that has to do with what we, as fallible human beings, should do. For instance, if a theory of ideal agents says that they should attend to all of the logical consequences of their beliefs, but we as human beings cannot do that, then the recommendations applicable to ideal agents are simply not applicable to us. *We* should do something else. As I use the term "theory of rational action", it is about what we, and other resource bounded cognitive agents, should do. I want to know how, given our cognitive limitations, we should decide what actions to perform. In other words, I want a theory of *real rationality* as opposed to a theory of *ideal rationality*.

This distinction is widely recognized, but it often seems to be supposed that as philosophers our interest should be in ideal rationality. The rationality a human can achieve is mere "bounded rationality"—a crude approximation to ideal rationality. But surely we come to the study of rational decision making with an initial interest in how we, and agents like us, should make decisions. This is the notion of rationality that first interests us, and this is what I am calling "real rationality". We might try to illuminate real rationality by taking it to be some kind of approximation to ideal rationality, but still our original interest is in real rationality.

Although theories of ideal agents are not directly about how real agents should solve decision problems, a plausible suggestion is that the rules of rationality for real agents should be such that, as we increase the reasoning power of a real agent, insofar as it behaves rationally its behavior will approach that of an ideal rational agent in the limit. This is to take theories of ideal rationality to impose a constraint on theories of real rationality. We can make this suggestion more precise by distinguishing, as I have elsewhere (1986, 1995), between "justified" choices and "warranted" choices. A *justified* choice is one that a real agent could make given all of the reasoning it has performed *up to the present time* and without violating the constraints of rationality. A *warranted* choice is one that would be justified if the agent could complete all possibly relevant reasoning. Two characteristics of real agents make this distinction important. First, for any cognitively sophisticated agent, reasoning is non-terminating. There will never be a point at which the agent has completed all the reasoning that could possibly be relevant to a decision. But agents have to act. They cannot wait for the completion of a non-terminating process, so decisions must be made on the basis of the reasoning that has been done so far. In other words, real agents must act on the basis of justified choices rather than waiting until they know that a choice is warranted. Second, it is characteristic of the reasoning of a real agent that almost all of its conclusions are drawn *defeasibly*. That is, the reasoning to date can make the conclusion justified, but acquiring additional information or performing additional reasoning may rationally necessitate the agent's changing its mind.[3]

For an agent that reasons defeasibly, we can characterize a warranted choice as one that, at some stage of its reasoning, the agent could settle on and never subsequently have to change its mind no matter how much

[3] For the most part, it will be unimportant in this book exactly how defeasible reasoning works. I have, however, discussed it at length elsewhere. See my (1995, 2002), and Pollock and Cruz (1999).

additional reasoning it might perform. This can be made more precise by talking about "stages of reasoning". The agent starts from some initial epistemic situation, and then at each stage of reasoning it either draws a new conclusion or retracts a previous conclusion. A conclusion (or choice) is warranted iff there is a stage such that (1) it is justified at that stage, and (2) it remains justified at all subsequent stages of reasoning.[4]

The warranted choices are those an ideal agent that was able to perform all relevant reasoning would make on the basis of the information currently at its disposal. One might suppose that warranted choices are those we want an agent to make. The difficulty is that a real agent cannot complete all the reasoning that might possibly be relevant to a decision. As remarked above, reasoning is a non-terminating process. Eventually the agent has to act, so we cannot require that it act only on the basis of warranted choices. The most we can require is that the agent perform a "respectable amount" of reasoning, and then base its choice on that. So a real agent acts on the basis of justified choices that might not be warranted.

In some cases it would actually be irrational for a real agent to make the warranted choice. For instance, suppose *P* and *Q* are logically equivalent, but the agent has not yet performed enough reasoning to know this. Suppose the agent has good reason to accept a bet that *P* is true at 2:1 odds. Suppose that choice is not only justified, but also warranted. Suppose, however, the agent has no basis for assessing the probability of *Q*. That is, it has no justified beliefs about the probability of *Q*. Then it would be irrational for the agent to accept a bet that *Q* is true at 2:1 odds. That choice would not be justified. But it would be the warranted choice, because if the agent performed *enough* reasoning it would discover that *Q* is equivalent to *P* and hence has the same probability.

Theories of ideal agents are theories of warrant. It might be suggested that the behavior of an ideal agent is the target at which real agents should aim, and hence theories of real rationality can be evaluated in terms of whether they approach the correct theory of ideal rationality in the limit. More precisely, a theory of real rationality, viewed as a theory of justified choice, implies a theory of warrant. We can think of a theory of ideal rationality as a theory of what the correct theory of warrant should say. The suggestion would then be that a theory of justified choice (real rationality) is correct iff its implied theory of warrant describes the behavior of an ideal rational agent (given some theory of what ideal rationality requires).

For epistemic cognition, real rationality and ideal rationality might be related in some such fashion, but it will turn out that there can be no such connection in the case of practical cognition. The set of justified choices will only converge to the set of warranted choices if there are always warranted choices to be made. But it will emerge in chapter 10 that there may often be no warranted choices for real agents living in the real world. It could be that no matter how good a solution the agent finds for a decision problem,

[4] There are two different concepts of warrant here. For a discussion of their interconnections, see chapter 3 of my (1995).

given enough time to reason there is always a better solution to be found. I will argue that this need not be an untoward result. The supposition that there must always be warranted choices turns on a misunderstanding of the logical structure of practical cognition—it assumes that decision problems always have optimal solutions. If they do not, then theories of warrant would seem to be irrelevant to theories of justified decision making.

So our target is a theory of real rationality—a theory of how real agents, with all their cognitive limitations, should make decisions about how to act. A theory of ideal rationality might conceivably be relevant to the construction of such a theory, somehow imposing constraints on it, but a theory of ideal rationality by itself cannot solve the problem of producing a theory of real rationality.

## 3. Human Rationality and Generic Rationality

A theory of real rationality is a theory of how one should proceed in making decisions. We might put this by saying that our concept of rationality is a procedural concept. I have discussed procedural rationality at length elsewhere in connection with epistemic cognition.[5] An agent's cognitive architecture determines how the agent goes about performing various tasks. However, as remarked in section 1, the human cognitive architecture leaves us some leeway in how to perform many tasks. Various cognitive tasks are under our control to some degree, and a theory of rationality aims at telling us how we should perform those tasks.

It is the fact that we have control over our own cognition that makes it possible for us to behave irrationally. When we can choose how to perform a cognitive task, we can do the wrong thing, thereby proceeding irrationally. So, for example, we conclude that agents should not engage in wishful thinking or hasty generalization, but observe that, nevertheless, they sometimes do. It is interesting to inquire why humans are so constructed that it is possible for them to behave irrationally. Why aren't we built so that it is rigidly determined that we always do the right thing? Sometimes this is because having the power to control the course of our thinking makes us more efficient problem solvers. But that same power enables us to behave irrationally. For instance, one thing we have control over is what to think about. By enabling a cognitive agent to engage in practical cognition about what to think about we enable the agent to focus on problems that it is more apt to be able to solve and to try to solve them in ways it thinks are more likely to be successful. But this same power to control what it thinks about enables an agent to *avoid* thinking about something. In particular, if the agent has a favored theory but has reason for suspecting that some particular consideration may constitute a problem for the theory, the agent

---

[5] I introduced the concept of procedural rationality in my (1986), specifically in connection with epistemic justification.

can avoid thinking about the possible problem—a classical instance of irrationality.

If we have the ability to do it wrong, what is it that determines when we are doing it right? That is, what makes rational cognition rational? It is a striking fact about human beings that we often find it easy to detect irrational cognitive behavior. How do we do that? Philosophers sometimes speak vaguely of their "philosophical intuitions", but that is to do no more than label the ability. When we catch an agent in irrationality, we know how to perform the task at hand. Knowing how to perform a task consists of knowing what to do as the task unfolds. We detect irrationality by knowing what to do and seeing that the agent does something different.

Knowing how to do something constitutes having *procedural knowledge*. I have many kinds of procedural knowledge. I know how to ride a bicycle, how to do long division, how to speak English, and how to engage in various kinds of epistemic and practical cognition. Having procedural knowledge for how to do something does not dictate that I always do it that way. Sometimes I fall off my mountain bike, make mathematical mistakes, speak ungrammatically, and reason incorrectly. When I know how to do something, I have either a learned or a built-in routine for doing it, but I do not always do it in that way. An important fact about human beings is that we have some ability to detect cases in which we do not conform to our learned or built-in routine. No doubt the functional explanation for this ability is that it enables us to correct our behavior and bring it into conformance with the way we know how to do things.

Because we do often try to bring our behavior into conformance with our procedural knowledge of how to do things, we can regard the learned or built-in routines as playing a normative role of sorts. As such, we can describe the routine in terms of a set of *norms*—rules for how to perform the routine. Chomsky (1957) introduced the *competence/performance distinction* as a way of talking about this. A performance theory regarding some activity is a theory of how people actually perform it. A competence theory is a theory about how they perform it when they are conforming to their procedural knowledge of how to perform it. So a competence theory is, in effect, a description of the procedural norms governing the way people have learned to perform the activity (or the built-on procedural norms if they are not learned).

Chomsky was interested in understanding what theories of grammar are about. His suggestion was that theories of grammar are competence theories of certain aspects of linguistic performance. Speakers of a language know how to speak grammatically, but they do not always do so. Because speakers often speak ungrammatically, a theory of grammar cannot be regarded as a performance theory. However, speakers have the ability to detect when they are diverging from their grammatical norms. Linguists assess grammaticality by asking speakers (or themselves) whether they regard particular sentences as grammatical. When they do this, they say that they are appealing to the "linguistic intuitions" of the language user. On Chomsky's account, these linguistic intuitions are just an exercise of speakers' ability to tell whether they are conforming to their procedural knowledge when they utter a particular sentence.

Chomsky's account of theories of grammar is now generally accepted in linguistics. In my (1986) I suggested an analogous account of epistemological theories. We know how to perform various cognitive tasks, among them being various kinds of epistemic cognition. Let us take *epistemic norms* to be the norms describing this procedural knowledge. Having this procedural knowledge carries with it the ability to detect when we are not conforming to our procedural norms. My suggestion is that the best way of understanding our epistemological intuitions is to take them as analogous to linguistic intuitions. That is, they are just a reflection of our ability to detect divergences from our epistemic norms. So an epistemological theory is a competence theory of epistemic cognition.

I propose that we extend this account to rationality in general. That is, a theory of rational cognition is a competence theory of human cognition. It describes our norms for how to cognize. I presume that our basic knowledge of how to cognize is built-in rather than learned. It is hard to see how we could learn it without already being able to cognize. So the basic norms for rational cognition are descriptive of our built-in cognitive architecture. More specifically, they are descriptive of those aspects of our cognitive architecture that guide our cognitive performance without rigidly determining it. They are descriptive of the norms provided by our cognitive architecture for how to perform those cognitive tasks over which we have deliberate control.

My reason for adopting this view of human rationality is that it seems to be the best way of making sense of the kind of support that philosophers typically offer for their claims about rationality. They appeal to their philosophical intuitions, but those are utterly mysterious unless we identify them with the familiar ability to monitor our own conformance to our procedural norms.

To summarize my conclusions so far, a competence/performance distinction arises for an agent whose cognitive architecture imposes rules for correct cognition but also enables the agent to violate them. A competence theory is a theory about performances that conform to the rules, and a performance theory is a general theory describing the agent's performance both when it does and when it does not conform to the rules for correct cognition. One way to think of a theory of rationality is as a theory of how to perform cognitive tasks "correctly", that is, in terms of the built-in rules of the cognitive architecture. This is to identify the theory of rationality with a competence theory of cognition. I will use the term "human rationality" to refer to a competence theory of human cognition. This approach generates a concept of rationality that is tightly tied to the details of the human cognitive architecture.

Although I take the preceding to be descriptive of standard philosophical methodology in investigating rationality, it is often useful to take a wider view of rationality, approaching it from the "design stance". We can ask how one might build a cognitive agent that is capable of satisfying various design goals. This immediately raises the thorny issue of what we should take to be the design goals of human cognition. But it turns out that by approaching cognition from the design stance we can explain many of the more general features of human cognition without saying anything precise about the design goals. For example, for a very wide range of design goals an agent will work better if it is capable of defeasible reasoning, if it treats perceptual input defeasibly, if it is able to reason inductively, if it is able to engage in long range planning, and so on. This generates a "generic" concept of rationality in which we are interested in how cognition might work in cognitive architectures aimed at a broad range of design goals. From this perspective, fine details of human cognition can often be viewed as fairly arbitrary choices in designing a system of cognition. For example, in building an agent, we may want to equip it with a set of rules sufficient for reasoning in the propositional calculus. There are many different sets of inference rules that will suffice for this purpose, and there may be little reason to choose some over others. Thus an arbitrary decision must be made. There is considerable psychological evidence to indicate that modus tollens is not among the built-in inference rules in human beings—it must be learned (Wason 1996; Cheng and Holyoak 1985). Thus from the perspective of human rationality, reasoning in accordance with modus tollens (before learning it through experience) is irrational. But there would be nothing wrong with building a cognitive agent in which modus tollens is built in. Relative to that agent's cognitive architecture, reasoning in accordance with modus tollens prior to learning about it from experience is perfectly rational.

I am primarily interested in understanding rational decision making in human beings. This makes it relative to the human cognitive architecture. However, those details of human cognition that could easily have been otherwise are of less interest to me than those that could not have been changed without adopting a radically different architecture. Thus in studying rational decision making we can ask two kinds of questions. We can ask how a human being should go about making a decision given the cognitive architecture that nature has endowed him with. But we can also evaluate the architecture itself, asking whether it could be significantly improved in various ways without radically altering the general form of the architecture. This second kind of question can in turn have implications for the first kind of question, because as noted above, although we cannot alter our built-in architecture, we often have the ability to employ learned behaviors to override built-in behaviors. Thus if there are better ways to solve decision problems than those dictated by our built-in procedures, we may be able to employ them. Our built-in procedures are often just default procedures, to be employed until we find something better, and when we do find better procedures our built-in architecture itself dictates using them to override the default procedures.

As we proceed, it will be very useful to keep in mind the distinction between evaluating a decision and evaluating a cognitive architecture. Theories of ideal rationality that cannot plausibly be adopted as theories about how real agents ought to make decisions may nevertheless be relevant to the evaluation of cognitive architectures. I will also argue, in chapter 3, that in at least one respect, human "evaluative cognition" is based upon a rather crude solution to the design problems it aims to solve. We cannot say that humans are irrational for working in the way they do. They cannot help the way they are built. But it is interesting to ask whether we could build a better agent. When the time comes, I will raise this issue again.

## 4. Decision Making

Before beginning the investigation of how decisions rationally ought to be made, it will be useful to reflect on what goes on in actual decision making. In a particularly simple case, I may just be deciding whether to perform some action *A*. For instance, I may be deciding whether to order the southwestern quiche for lunch. This often involves comparing *A* to a number of other alternatives. For instance, should I instead order the chicken salad sandwich? In a particularly simple case, my choice could just be between *A*-ing and not *A*-ing.

It is important to realize that decisions are always made in advance of acting. You cannot literally decide to do something *now*. If by "now" you mean "at this very moment", then either you are already performing the action or you are not performing the action. It is too late to decide. Of course, your decision might be about what to do within the next second or two. But we often have to make decisions far in advance of the time they are to be carried out. This is for at least three reasons. First, I may have to do other things before I can carry out a decision. For instance, if I decide to paint my bedroom, I may have to buy the paint. Second, decisions can involve a whole course of actions rather than a single action. I may decide to paint two rooms, doing the bedroom last. The decision has to be made early enough that I can paint the first room before painting the bedroom, and hence the painting of the bedroom may not occur until some time after the decision is made. When we decide to perform a whole sequence of actions, we are adopting a *plan*. I will say much more about plans over the course of this book.

The third reason decision making often precedes acting by an extended period of time is that decision making can be difficult, consuming considerable cognitive resources and taking quite a bit of time. In the course of making a decision we may have to acquire additional information and we may have to think long and hard about it. We may not have time do all this just shortly before the time to act. We may have to do it well in advance. This is particularly common when we have to perform a number of actions in quick succession. Consider planning a driving route through an unfamiliar city on busy highways. You must plan ahead, memorize where you will go at each intersection, and then follow your plan without further deliberation.

It is because we are resource-bounded cognitive agents that we must often plan well in advance of acting. Having chosen a plan—made a decision—our default procedure must be to follow it automatically. However, if things do not go as expected, we must be able to reopen deliberation and reconsider the plan. For instance, if you run into unexpectedly high traffic on your chosen route through the city, you must be able to consider changing your plan.

A further complication is that when we plan ahead we will be subject to varying degrees of ignorance about the conditions under which the plan will be executed. For example, I do not normally plan ahead about which traffic lane to use on a particular leg of my route. I decide that in light of the flow of traffic around me as I drive. Decisions about the details of my plans are often best left until the last minute when I will know more about the circumstances in which I am executing the plan. To accommodate this, plans are typically somewhat schematic.[6] I adopt a skeletal plan far in advance, and then slowly fill in the details by making further decisions as the time for acting draws nearer. The end result of such temporally extended planning is a better plan. The resulting plan is not the result of a single act of decision making. It results from a temporally extended sequence of decisions.

There is another reason for adopting skeletal plans. I may have to make a decision before I have time to work out all the details. For example, if I am invited to give a talk at a conference in Copenhagen nine months hence, I may have to decide quickly whether to do it, without having the time to plan exactly what flights I will take. I can work out the details later when I have more time and a lighter cognitive load.

The upshot is that our decisions result in our adopting plans that are schematic to varying degrees. As the time to act draws nearer, planning continues, filling in more details. Note that I may start executing the first part of my plan before filling in the details of the later parts. For instance, I start driving through the city before I decide what traffic lanes to use when I am on the far side of the city.

It is noteworthy that our plans almost never involve precise specifications of when we are going to perform the actions prescribed by the plans. That is not determined until we actually do it. For example, I might decide to buy milk at a convenience store, so I go to the store, take a carton of milk out of the cooler, take it to the cashier, and pay her for it. But I do not decide beforehand at precisely what instant I will hand the money to the cashier. That is not determined until I do it. Furthermore, when I actually do hand the money to the cashier, that does not seem to be a matter of deciding. We certainly do not deliberate about when to do it at the instant we do it. If we were still deliberating, we would not be ready to do it yet, and if we are doing it we must have stopped deliberating at some earlier time. There has to be some point at which deliberation ends and automatic processing takes over. The initiation of the action must be an automatic process rather than one we do deliberately. When we are through deliberating, the action goes on a queue of actions waiting to be performed, and actions are initiated in the order they are retrieved from the queue, without the cognizer thinking about it any further. Philosophers sometimes claim that

[6] This point was emphasized in my (1995).

actions are initiated by a mental act of "willing", but I am not sure what that is supposed to amount to. Do we have to will ourselves to will? If so, we seem to be threatened with an infinite regress. On the other hand, if we can initiate a willing without willing to will, why can't we initiate a finger wiggling without willing that? I think that the philosophers who appeal to willing are over-intellectualizing action initiation. Once I decide to perform the action, and put it on the queue, my action is initiated by my cognitive system, not by me.

One consequence of the schematicity of plans is that I may adopt a plan—form the intention to execute it—but fail to execute it despite the fact that I do not change my mind. For example, I may decide to go to the grocery store this afternoon. But at the time I make this decision I do not decide precisely when to go—just sometime this afternoon. The afternoon passes and I am always busy doing other things, with the result that I do not get to the store. But I did not change my mind about going.

The general picture that emerges from this is that we deliberate and decide on either individual actions or entire plans. Forming an intention amounts to deciding to perform an action or adopt a plan. Adopting a plan has the consequence that, unless we have a lot of cognitive free time, we act on the plan without reconsidering it unless we acquire new information that forces us to reconsider it. On the other hand, we *can* always reconsider our decisions, but as long as we have a pretty good plan, our limited cognitive resources will normally be expended on other more pressing matters.

Although we do not usually reconsider our plans once we have adopted them, execution cannot be entirely automatic because further decision making will be required to fill in the details. It can happen that we are unable to find a good way of filling in the details, in which case the plan will be aborted. For example, when I go to pay for the milk, there may be a power outage that shuts down the cash register, leaving me without a way of paying.

Armed with this understanding of what goes on in decision making, our objective is to investigate what constraints rationality places on the process. How should a cognitive agent, subject to realistic cognitive limitations, decide what to do?

## 5. Classical Decision Theory and the Optimality Prescription

Throughout contemporary philosophy and cognitive science, one encounters the almost universal presumption that the problem of rational choice is essentially solved by classical decision theory. One of the main conclusions of this book will be that classical decision theory is wrong—when allowed its head, it leads to intuitively incorrect prescriptions about how to act. There is something right about classical decision theory, but the problem of constructing a theory of practical cognition becomes that of replacing classical decision theory with a more sophisticated theory that retains its insights while avoiding its shortcomings.

The fundamental prescription of classical decision theory is the *optimality prescription*, according to which, when one is deciding what to do, rationality dictates choosing the alternative having the highest expected value. This principle provides the cornerstone on which theories of subjective probability are constructed, underlies so-called belief/desire psychology, and drives most work on practical reasoning. It plays a pervasive role in contemporary philosophy, and has rarely been questioned. But I will argue that the principle is false, for several different, essentially orthogonal, reasons.

By "classical decision theory" I mean the nexus of ideas stemming in part from Ramsey (1926), von Neumann and Morgenstern (1944), Savage (1954), Jeffrey (1965), and others who have generalized and expanded upon it. The different formulations look very different, but the basic prescription of classical decision theory can be stated simply. We assume that our task is to choose an action from a set **A** of *alternative actions*. The actions are to be evaluated in terms of their outcomes. We assume that the *possible outcomes* of performing these actions are partitioned into a set **O** of pairwise exclusive and jointly exhaustive outcomes. In other words, it is logically impossible for two different members of **O** to both occur, and it is logically necessary that some member of **O** will occur. We further assume that we know the probability $\text{PROB}(O/A)$ of each outcome conditional on the performance of each action. Finally, we assume a *utility measure* $\mathbf{U}(O)$ assigning a numerical "utility value" to each possible outcome. The *expected value* of an action is defined to be a weighted average of the values of the outcomes, discounting each by the probability of that outcome occurring if the action is performed:

$$\mathbf{EV}(A) = \sum_{O \in \mathbf{O}} \mathbf{U}(O) \cdot \text{PROB}(O/A).$$

The crux of classical decision theory is that actions are to be compared in terms of their expected values, and rationality dictates choosing an action that is *optimal*, that is, one such that no alternative has a higher expected value. This is what I am calling "the optimality prescription". To illustrate, suppose we are comparing two actions. We can push button 1, or we can push button 2. If you push button 1, there is then a probability of 1/3 that you will receive \$3, and a probability of 2/3 that you will receive \$6. If you push button 2, there is then a probability of 1/2 that you will receive \$2, and a probability of 1/2 that you will receive \$7. Which button should you push? Computing the expected values:

$$\mathbf{EV}(\text{button 1}) = 3/3 + (6 \times 2)/3 = 5$$

$$\mathbf{EV}(\text{button 2}) = 2/2 + 7/2 = 4.5$$

So the optimality prescription recommends pushing button 1.

Now I turn to some technical details that can be skipped without loss of comprehension. Throughout this book I will isolate such technical material by displaying it in a different font, as below.

For the technically inclined, it can be noted that it is sometimes necessary to adopt a more complex definition of expected value. The above definition assumes that we need only distinguish between finitely many possible outcomes. This finiteness assumption can fail in some decision problems. For example, we might want to distinguish amounts of some resource that is produced as a result of performing an action. If the amounts are measured by real numbers, and more is better, then for every real number *r*, the outcome of producing a quantity of measure *r* has to be regarded as a distinct possible outcome. This complication is handled by replacing the definition of expected value as a finite sum by a definition in terms of integrals. There may be an infinite set $\mathcal{W}$ of possible outcomes ("possible worlds"). The classical theory is extended to accommodate this possibility. If we have a probability distribution over the values of the worlds, we can generalize the definition of expected value as follows:

$$\mathbf{EV}(A) = \int_{-\infty}^{\infty} r \cdot \frac{d}{dr}\mathbf{PROB}(\mathbf{U}(w) \leq r \,/\, w \in \mathcal{W} \;\&\; A \text{ is performed in } w)dr$$

where *w* is a random variable ranging over possible worlds in which *A* is performed. Technically, the integral characterizes the mathematical expectation of the function **U**(*w*) on the condition "$w \in \mathcal{W}$ & *A* is performed in *w*". If there are just finitely many outcomes, the original characterization of expected values as finite sums can be derived from the integral formulation.

The intuitive idea of a mathematical expectation should be clear to all. It is just a weighted average of values, where values are weighted by the probability of their occurring. To express this generally, we must use integrals. However, I realize that many of my readers will be uncomfortable with integrals. I will adopt a notation for mathematical expectations that conceals the integrals and lets readers use their intuitions. The mathematical expectation of a function *f*(*x*) over all the *x*'s satisfying a condition φ is a weighted average of the values of *f*(*x*) for *x*'s satisfying φ, and I will henceforth abbreviate it as follows:

$$\mathsf{EXP}(f(x)/\varphi x) = \int_{-\infty}^{\infty} r \cdot \frac{d}{dr}\mathbf{PROB}(f(x) \leq r \,/\, \varphi x)\, dr$$

Note that the EXP notation has the effect of concealing the reference to the probability, and that can be dangerous. Over the course of the book, we will distinguish between a number of different kinds of probability, and any of those can be employed in defining mathematical expectations. So when using the EXP notation, it will be important to be clear about what kind of probability is being employed.

For present purposes, we can write

$$\mathbf{EV}(A) = \mathsf{EXP}(\mathbf{U}(w) \,/\, w \in \mathcal{W} \;\&\; A \text{ is performed in } w).$$

The kind of probability employed here is whatever kind of probability enters into decision-theoretic reasoning. At this point we are being intentionally vague about that. Eventually, I will urge that we should be employing a variety of causal probability defined in chapter 8.

In talking about expectations, we must use integrals for generality. Fortunately, for understanding the philosophical significance of the mathematical expectations of various quantities, we can generally ignore the fact that they are defined using integrals. When there is some point to expanding the definitions, we can typically ignore the general case and just consider the finite case where expectations can be represented as finite sums. In particular, I will typically assume that expected values can be defined as finite sums.

It is the optimality prescription that constitutes the meat of classical decision theory as a theory of rational choice. Different approaches to decision theory arrive at the optimality prescription in different ways. Savage lays down axioms governing rational choice, and derives the optimality prescription from the axioms. Von Neumann and Morgenstern lay down axioms governing preference between consequences, where the consequences include participation in lotteries. Jeffrey lays down axioms regarding preferences between propositions, and then identifies actions with those propositions one can make true. These differences are important, but what I want to focus on here is the optimality prescription itself. It is common to all the different versions of classical decision theory.

It is important to notice that, at least as formulated above, classical decision theory does not include in account of how the set **A** of alternative actions is to be chosen. For the purposes of the optimality prescription, these are taken to be given, but for application to real decision making, we must know to which alternatives a given action should be compared. This will become a crucial issue as the book proceeds. Decision theorists typically dodge this problem by defining a *decision problem* to be a certain kind of algebraic structure that includes a specification of **A**, **O**, and the probabilities. Thus "decision problem" becomes a technical term. Decision theory supposes that we must first "formulate a decision problem", and then decision making is done against the backdrop of that formulation. Decision theorists usually give little consideration to the question of how to cast our decision making in the form of a decision problem.[7] It is important to realize that real decision problems do not come to us in this nicely packaged form. Real agents are faced with the question, "Given what I believe about the world, should I do *A*?" or more generally, "What should I do?" There is no presumption that some particular set of actions **A** includes all of the relevant possibilities or that some particular set **O** includes all the relevant possible outcomes. Part of the task of rational choice is to decide what actions and outcomes to consider, and a complete theory must include an account of how to do that. One of the most important conclusions of this book will be that if we take this problem seriously, we will find that rational decision making cannot usually be cast as a decision problem in quite the way assumed by classical

[7] An exception is Joyce (1998).

decision theory.

It is my conviction that the optimality prescription is wrong for several essentially orthogonal reasons. The book will formulate these objections, and make a series of modifications to the theory that are aimed at meeting them. To forestall an objection to my own account, let me emphasize that the target of my criticisms is not decision theory per se, but the optimality prescription when construed as a theory of rational choice between ordinary actions by real agents. This is what I mean by *classical* decision theory. When I cast aspersions on decision theory, this is my target. This is certainly the way most philosophers ordinarily understand decision theory. However, what we might call "the discipline of decision theory" has evolved over the years, partly in response to some of the difficulties I will discuss. For example, a way of trying to meet the objections of chapter 10 is to reconstrue decision-theoretic actions as plans, and a lot of work in decision theory has gone in that direction. This is to move away from what I am calling "classical decision theory". I will argue that this still does not generate a correct theory of rational choice, but it is an important move along the road to one.

Before attacking the optimality prescription, it will be useful to consider more carefully just how it is to be understood. It proceeds in terms of expected values, and they are defined in terms of probabilities and utilities. Specifically, the definition requires a way of measuring both probabilities and utilities in terms of real numbers. So the book will be divided into three parts. Part I, consisting of chapters 2–5, will take up the question of where this real-valued measure of utilities comes from. Part II, consisting of chapters 6–8, will investigate the nature of the probabilities required. Part III, consisting of chapters 9–12, will investigate how to use these values and probabilities in decision making. For readers with specialized interests, the parts can be read more or less independently.

The criticism of the optimality prescription is spread over chapters 8, 9, and 10. Chapter 8 discusses the familiar difficulties originating from Newcomb's problem (Nozick 1969). Those problems gave rise to "causal decision theory" (see Gibbard and Harper 1978; Lewis 1981; Skyrms 1982; Sobel 1978), which differs from classical decision theory in the manner in which probabilities enter into the definition of expected value. I will propose a definition of causal probability that, when used in the otherwise standard definition of expected value, avoids those problems.

Chapter 9 raises a problem that has been almost entirely overlooked. This is that in deciding on a course of action, a decision maker will often not know with certainty which actions he will be able to perform. A theory of rational choice ought to take this into account. I will argue that this necessitates further changes to the way expected values are defined.

Chapters 8 and 9 discuss difficulties that are repairable without changing either the general form of classical decision theory or the optimality prescription. Chapter 10 raises a much deeper difficulty that necessitates changing the entire framework of decision theory and the theory of rational choice. Classical decision theory focuses on choosing individual actions, but as was illustrated above in section 4, our decisions are often to adopt entire plans.

Classical decision theory assumes that a plan can be adopted by focusing on the individual actions making up the plan. One of the most important conclusions of this book will be that this reduction does not work. In deciding what to do, actions cannot be evaluated in isolation—they must be evaluated in the context of other actions that may interact with them either constructively or destructively—that is, they must be evaluated as parts of plans. To handle this, the natural suggestion is that the optimality prescription should be applied to plans rather than individual actions. But it is argued in chapter 10 that there is no way to characterize the set of alternatives to a plan in such a way that there will normally be an optimal alternative having a maximal expected value. Furthermore, in those unusual cases in which there is an optimal choice, it will be computationally impossible for a real cognizer to reliably find it. Finally, it will be argued that plans cannot be meaningfully compared in terms of their expected values anyway. Chapters 11 and 12 explore some of the details of a theory of practical cognition that accommodates these difficulties.

# Part I

# Values

# 2
# Evaluative Cognition and the Evaluative Database

## 1. The Doxastic/Conative Loop

Recall the doxastic/conative loop from chapter 1. Cognitive agents form beliefs representing the world, evaluate the world as represented, construct plans for making the world more to their liking, and perform actions executing the plans. Evaluation plays a pivotal role in linking beliefs about the world with rational decision making. This chapter and the three following will focus on evaluation as a cognitive enterprise performed by cognitive agents. I am interested in both how it is performed in human beings and how it might be performed in other cognitive agents.

Evaluative cognition produces assessments of value. However, this may or may not be the same concept of value that is the focus of interest in value theory. Value-theoretic investigations are predominantly metaphysical. The present chapter pursues the epistemology and cognitive science of value rather than its metaphysics. More specifically, this investigation is concerned with those aspects of rational cognition that are used in the comparative assessment of competing actions or plans, and for present purposes *value* is simply defined to be whatever such assessments measure. Thus rather than starting from a metaphysics of value and asking how we can learn about values, I start with an investigation of certain aspects of cognition. Where do cognitive agents get the evaluations they employ in practical cognition? How are values computed? It turns out that purely computational considerations can take us a long way towards answering these questions. The metaphysically inclined can go on to ask about the nature of the values that are the objects of such cognition and how they are related to other philosophical enterprises, but this book will not pursue those matters.

The main use of values in practical cognition appears to be in computing the expected values of actions. This requires a real-valued measure of values—a "utility measure"—and it involves multiplying utilities and probabilities and summing the results. For that to make sense, the utility measure must be a cardinal measure of value, not just an ordinal measure. Ordinal measures provide only a rank ordering of whatever it is that is being measured. Cardinal measures purport to actually measure the size of a quantity. The mark of a cardinal measure is that one can meaningfully add measures or multiply measures by a real number to get the measure of a larger quantity. The definition of expected value presupposes the ability to perform mathematical manipulations on the measures of utilities, multiplying them by probabilities and adding the results, so a cardinal measure is required. In other words, larger numbers do not just tell us that one outcome is better than another—they tell us how much better. Any theory of rational choice based on expected values must face the problem of telling us where these numbers come from.

Furthermore, it is the agent that is deciding what plans to adopt, so it is not enough for the cardinal measure of value to simply exist—the agent must have cognitive access to it in order to perform decision-theoretic evaluations of alternatives. In other words, the agent must be able to compute utilities in a way that makes decision-theoretic evaluations possible. I will ask how that can be done in general (in any cognitive agent), and more specifically how it is done in human beings.

A common answer, suggested by work in subjective expected utility theory, is that evaluative cognition begins with binary preferences. However, In this chapter I will argue that it is computationally impossible for a database of binary preferences to represent the fundamental evaluative data structure in a sophisticated cognitive agent. The only way such an agent can function is by storing a small subset of values, and then computing others in terms of the stored values. That requires something like a real-valued cardinal measure of values. But this seems to conflict with the fact that humans cannot introspect real numbers as measures for their values. This will be resolved below by arguing that humans accomplish the same thing by employing "analog representations". The use of analog representations of quantities is a pervasive feature of human cognition.

## 2. Preference Rankings

The question is, where does a cognitive agent get the measures of values that are required for computing expected values and applying the optimality prescription? Let us begin by considering an answer that is suggested by subjective expected utility theory (discussed in chapter 5) and underlies much of modern economics. This answer begins with the observation that although decision theory requires a cardinal measure of value, human beings are unable to assign numbers to values simply by introspecting. On the other hand, humans can introspect *preferences*. That is, they can tell introspectively that they prefer one type of situation to another.

The values needed for decision-theoretic reasoning are the values of *outcomes*. These are "types of situations". It is useful to make a type/token distinction here. *Situation tokens* are "total" ways the world might be, that is, complete possible worlds. Practical cognition aims at changing the actual world so that it exemplifies a situation-type the agent values. *Situation-types* abstract from the situation tokens and are partial descriptions of ways the world might be. Such partial descriptions describe the situation in terms of some of its features, so I will refer to situation-types more compactly as *features*. Values attach to features, and if they are to be derived from preferences, the preferences must also be between features.

Assuming that our preferences are transitive, we get a preference ranking of all the items between which we have preferences. Frank Ramsey (1926), John von Neumann and Oscar Morgenstern (1944), and Leonard Savage (1954) showed in different ways that if our preference ranking includes comparisons of wagers, and the ranking satisfies some plausible axioms, then it is possible to generate a cardinal measure supporting decision-theoretic reasoning. Theorems to this effect are known as "representation theorems", and subsequent authors (Bolker 1966; Armendt 1986, 1988; Joyce 1998) have established a variety of representation theorems that employ different assumptions about the preference ranking.

Representation theorems are thought to be a very important part of decision theory. For example, Joyce (1998, p. 224) writes, "No decision theory is complete until it has been supplemented with a representation theorem that shows how its 'global' requirement to maximize expected utility will be reflected at the 'local' level as constraints on individual beliefs and desires." This is a very puzzling claim. It is not at all clear what the philosophical point of these representation theorems is supposed to be. If they are supposed to justify the optimality prescription, they fail because all existing representation theorems require suppositions about preferences that have much less claim to intuitive obviousness than does the optimality prescription itself.

Perhaps the point of representation theorems is to meet the charge that as we are unable to introspect numerical measures of values, there is no reason to think that such a cardinal measure exists. That is a serious problem, and if the preference axioms were obviously true then the representation theorems would accomplish something important. But everyone seems to agree that they are less than obvious. Furthermore, it will be argued over the course of this book that the optimality prescription, as thus far formulated, is incorrect, so a representation theorem deriving it from preference axioms would really only show that the preference axioms are false. In section 3 I will suggest an alternative way of justifying the assumption that there is a cardinal measure of values, so representation theorems will not be needed for that purpose.

Regardless of what else they may establish, it must be emphasized that representation theorems show nothing directly about how human beings perform practical cognition. In particular, they clearly do not show that humans proceed by recovering a cardinal measure from their preference rankings and then use it to reason decision-theoretically. Despite this, I have found that a huge number of philosophers, economists, decision analysts, psychologists, and other cognitive scientists believe that what is basic to evaluative cognition is a set of binary preferences, and all other aspects of evaluative cognition must derive from these preferences. In philosophy this is often taken as an explication of the so-called Humean theory of desire, according to which desires are basic and uncriticizable (except by reference to other desires).

Let us ask a more general question. Regardless of how humans work, *could there be* rational agents that work in this manner? The proposal would be that the fundamental value-theoretic data structure to which the agents appeal is a preference ranking. I will now argue that this is impossible. An agent could not be built that works this way in a complex environment. To see this, consider how many features would have to be included in the preference ranking. Perhaps every feature should be included. How many features is this? Features are types of situations, so we might plausibly suppose that every set of possible states of the world determines one feature. Then if the number of possible states of the world is $N$, the number of possible features is $2^N$.

To estimate the number of possible states of the world, it was estimated a few years ago that there are $10^{78}$ elementary particles in the universe. If we take the state of a particle to be determined by four quantum states each having two possible values (a gross underestimation), each particle can be in 16 states, and so there are $16^{10^{78}}$ states of the universe and $2^{16^{10^{78}}}$ features. This is a bigger number than we can write in non-exponential form. It would contain more digits than the number of elementary particles in the universe. Clearly, a cognitive agent cannot contain a data structure constituting an explicit preference ranking of this many features.

It is fairly obvious that, as large as the preceding number is, it is a considerable underestimation of the true number of features that can be exemplified in the real world. Some of the parameters describing elementary particles, like position, are not two-valued but continuous-valued. This means that there are infinitely many possible states of the world, and correspondingly infinitely many features. So a preference ranking comparing all pairs of features would have to be infinite. Obviously, a real agent cannot contain an infinite data structure.

Apparently human beings do not explicitly store preference rankings including all features. Perhaps, somehow, we are able to ignore most features and just rank those that are significant to us. These will be the features characterized by parameters that affect things that really matter to us. Suppose we could confine our attention to 300 two-valued parameters. Then there will be 600 "simple" features each consisting of one of these parameters having a specific value. All other features will correspond to conjunctions of these simple features. This results in there being $2^{600}$ complex features. Many of these will be inconsistent, containing both a simple feature and its negation, and we can reasonably preclude them from the ranking. But $2^{300}$ of them will be consistent. $2^{300}$ is approximately $10^{90}$. To appreciate how large a number this is, notice that it is 12 orders of magnitude larger than the estimated number of elementary particles in the universe. This is an absurdly large number. Once again, real agents cannot store an explicit preference ranking of this many features, and this is what we get by ignoring all but 300 two-valued parameters. Surely, in the real world, more than 300 parameters are relevant to the values of features.

Perhaps we do not have to include all features in a preference ranking. It seems reasonable to propose that the preference ranking need only include features that have nonzero value. An agent is apt to be indifferent to most features, so this seems to produce a markedly smaller preference ranking. But it is still not small enough. Let *P* and *Q* be features, where *P* is "value-laden", that is, has a nonzero value, and *Q* is not. Then *P* will be included in the preference ranking and *Q* will not. However, unless *Q* interacts with *P* in such a way as to cancel its value, (*P*&*Q*) will also be value-laden, and so if all value-laden features are included in the preference ranking, (*P*&*Q*) must be included. Hence little is gained by leaving *Q* out of the preference ranking. To illustrate, suppose again (unrealistically) that states of the world can be characterized by just 300 two-valued parameters. Suppose (even more unrealistically) that just 60 of these simple features are value-laden, and suppose that compound features are value-laden only by virtue of containing one or more value-laden simple features as constituents. There will be $2^{300}$ consistent complex features, but they need not all be ranked. However, the only features that need not be ranked are those containing no value-laden constituents. There will be $2^{240}$ of these. $2^{240}/2^{300} = 2^{-60} =$ .000000000000000008. Thus only a minuscule proportion of the features are omitted from the ranking.

Perhaps we can simplify the preference ranking still further. Presumably it will usually be true that if *P* is value-laden and *Q* is not, then the value of (*P*&*Q*) will be the same as that of *P*. In that case we can simply leave (*P*&*Q*) out of the ranking, and when the time comes to compare it with other features we *compute* a preference by taking it to be the same as the preference for *P*. We cannot always omit conjunctions (*P*&*Q*) from the preference ranking, because sometimes *P* and *Q* will interact in such a way that the value of (*P*&*Q*) is different from that of *P*, but we can record that fact by including (*P*&*Q*) in the preference ranking. Using this strategy, when a conjunction is not contained in the preference ranking and we want to compute a preference between it and some other feature, we can do that by identifying its place in the ranking with that of the largest conjunction of a subset of its conjuncts that is explicitly contained in the ranking. There may be more than one such conjunction, but they will all be ranked alike if the absence of the conjunction from the ranking means that it has the same value as any smaller conjunction from which it can be obtained by adding value-neutral conjuncts.

This strategy achieves a significant decrease in the size of the preference ranking. In the above example we would only have to include $2^{60}$ elements in the ranking. However, that is still a pretty large number, approximately $10^{18}$. The best current estimates are that this is larger than the entire storage capacity of the human brain,[1] and that is from just 60 value-laden simple features. Realistically, human beings must be faced with at least 300 value-laden simple features (and probably orders of magnitude more). That produces a preference ranking containing at least $2^{300}$ items, which is again 12 orders of magnitude larger than the number of particles in the universe.

---

[1] See Landauer (1986).

This cannot possibly be the way human beings record values.

Could we avoid this difficulty by allowing an agent to have an "incomplete" preference ranking that only ranks some combinations of value-laden features?[2] Not if it is to form a useful basis for decision making. The number of items in the preference ranking cannot be greater than the entire storage capacity of the human brain. Suppose that allows a preference ranking of $10^{18}$ items (probably a considerable over-estimation). If we suppose that all value-laden features of the world are combinations of just 300 simple value-laden features of the world (surely a gross underestimation), a preference ranking of $10^{18}$ items will include one out of every $10^{72}$ combinations of value-laden features. Presumably, some of the omitted combinations will never occur in the real world, but what possible reason could there be for expecting only one out of every $10^{72}$ of them to occur? Without a very strong argument, such a supposition seems simply silly. But then an agent with only $10^{18}$ items in its preference ranking will rarely have preferences between actual states of the world. This cannot be the basis for decision making.

It must be concluded that the fundamental value-theoretic data structure in human beings does not take the form of a preference-ranking. This is not to say that humans do not have the requisite preferences, but just that they are computed from something else that can be stored more compactly. What might that be? There is a simple answer—just store numerical assignments of value to value-laden features. Preferences can then be computed by comparing the numbers. But is this really more compact? If we have to store a numerical value for every value-laden feature, we are no better off than with preference rankings. Of course, we do not have to store numbers for *every* feature. Just as in the case of preference-rankings, we can take the absence of a feature from the database to signify that its value is the same as that of the largest conjunction of its conjuncts that is present in the database. But still, that makes the database no smaller than the smallest preference ranking we were able to produce above.

However, for a database of numbers, if the numbers represent a cardinal measure, a considerable additional simplification can be achieved. Just as we can assume defeasibly that conjoining a value-laden feature with a value-neutral one will produce a conjunction whose value is the same as that of the value-laden conjunct, so it seems reasonable to assume defeasibly that conjoining two value-laden "simple" features will produce a conjunction whose value is the sum of the values of its conjuncts. In that case, it can be omitted from the database and a value computed for it when needed. Of course, value-laden features are not always independent of one another. For example, suppose I want a dish of vanilla ice cream. A playful friend might offer to provide the ice cream if I will eat a dill pickle first. Eating the dill pickle interacts with eating the ice cream in such a way as to nullify its value. When we cannot compute the value of a conjunction by summing

---

[2] E.g., Kaplan (1996) proposes something like this.

the values of the conjuncts, we can record the lack of independence by including an assignment to the conjunction in the database. Let us say that *P* and *Q* are *value-theoretically independent* iff the value of (*P*&*Q*) is the sum of the value of *P* and the value of *Q*. On the assumption that simple features are usually value-theoretically independent, a database recording values produced by 300 value-laden simple features will require on the order of 600 entries, as opposed to the $2^{300}$ entries required in a preference ranking. This difference is the difference between the trivial and the impossible. Of course, this assumes that value-laden features are generally value-theoretically independent. This does not seem obviously wrong, but I have given no argument for it. However, this will be rectified in chapter 5, which defends the assumption in detail.

It is worth noting why the same simplification cannot be achieved using preference rankings. It might be supposed that if we can assume defeasibly that value-laden states are value-theoretically independent then we need not store most conjunctions in the preference ranking. Can't we compute the position of the conjunction from the position of the conjuncts? The answer is that we cannot. The preference ranking gives us only an ordinal measure of value. To compute the value of a conjunction from the values of its conjuncts, we have to be able to add values, which requires a cardinal measure. For example, suppose I prefer *A* to *B* and *C* to *D*. Should I prefer *A*&*D* to *B*&*C*? There is no way to tell just on the basis of the initial preferences. On the other hand, if I have a utility measure $\mathbf{U}(x)$, and *A*,*B*,*C*, and *D* are value-theoretically independent, then we can compute that *A*&*D* is preferable to *B*&*C* iff $\mathbf{U}(A) + \mathbf{U}(D) > \mathbf{U}(B) + \mathbf{U}(C)$.

The preceding discussion was predicated on the assumption that the simple value-laden features are produced by setting the values of two-valued parameters. At least some of the relevant parameters, like position, are continuous-valued. Position is probably not itself a value-laden feature, but when combined with other value-laden features it can change their values. An ice cream cone on the moon is not nearly so valuable as one in my hand. Continuous-valued parameters make the preference ranking infinite, in which case it clearly cannot constitute the evaluative data structure underlying the value computations of either human beings or artificial rational agents. On the other hand, continuous-valued parameters need create no difficulty for storing numerical values. Rather than storing a constant value, we simply store a function of the parameter. This will be discussed at greater length in chapter 3.

The upshot is that storing a cardinal measure of values for simple features is a vastly more efficient way of storing values than storing them in the form of a preference ranking. The cardinal measure allows us to use arithmetic in computing the values of composite features, and that in turn allows us to omit the computable values from the value-theoretic database. I will turn to the details of this computation in chapter 5, and the justification for the assumption that features are generally value-theoretically independent. But first let us consider whether it is psychologically realistic to suppose that humans do store cardinal measures of value.

## 3. Analog Representations of Values

It was remarked above that it is not sufficient for decision-theoretic reasoning that values simply *have* cardinal measures. The decision maker must have cognitive access to those measures in order to use them in computing expected values. If evaluative cognition makes use of a database of stored values, then it is plausible to suppose that a cognitive agent gets access to their stored measures by some sort of introspection. Basically, the agent introspects "how much it likes" a feature. This is the only obvious way to build a cognitive agent capable of decision-theoretic reasoning. But there is a problem with this suggestion. Human beings are not able to introspect numerical values for how much they like features, and they are quintessential general-purpose cognitive agents.

The observation that humans are not able to introspect numerical values for their conative states is responsible for the retreat to preference rankings in place of numerical measures of value. However, I argued in section 2 that preference rankings cannot provide the computational basis for evaluative cognition. This is not to deny that humans have preferences, but just to deny that the preferences constitute the basic evaluative database from which evaluative cognition proceeds. The only computationally feasible way of grounding evaluative cognition in an agent capable of functioning in a complex environment seems to be via a cardinal measure of value.

### *3.1 Analog Representations of Quantities*

The solution to this quandary lies in recognizing that numbers are not the only possible mental representations of quantities. Humans regularly employ and manipulate what might be called "analog representations" of quantities. For example, consider human judgements about length. We are quite good at comparing lines and judging which is longer. This might suggest that all we are really able to do perceptually is make an ordinal comparison, judging that one line is longer than another. This would be analogous to the claim that our fundamental access to values is via preferences. But in fact, we are able to make more elaborate comparisons of lengths. We can judge not only that one line is longer than another, but also that it is *much longer*, or perhaps *just a little longer*. Such judgments make no sense if we have only ordinal comparisons. We can even judge perceptually that one line is *twice as long* as another. For that to make sense, we must be employing something like a cardinal measure, even though we are not using numbers to represent the sizes. Instead we employ analog representations. Analog representations can be mapped onto numbers, and the numbers behave like a cardinal measure. It is common in psychometrics to construct a rough mapping of this sort by asking subjects to rate quantities on a scale of 1 to 10. This is often something that subjects are able to do fairly easily and with considerable consistency.

The ability to compare lengths perceptually enables us to construct rulers and introduce numerical measures of length, and as a result of making a mental comparison of a line with a remembered ruler one is often able to look at a line and make a numerical estimate of its length. But it is important to realize that the resulting number is not our primary representation of the length. We can judge that one line is twice as long as another without first constructing numerical estimates of their lengths, and unless we have reason to be particularly careful, that is what we generally do. We do not naturally think of lengths in terms of numbers.

My experience has been that philosophers find the above observations surprising, but psychologists who study "mathematical thought" in humans and other vertebrates find them commonplace and regard them as long established. Gallistel et al. (2000) summarize a large amount of data supporting this view of what we might call "quantitative cognition" in not only humans, but also rats, pigeons, and monkeys. The mathematical abilities of non-human vertebrates extend not only to cardinality judgments in finite sets, but also to judgments about continuous quantities like length, area, time, weight, and as we will see below, value and even expected value. They conclude, "In summary, research with vertebrates, some of which have not shared a common ancestor with man since before the rise of the dinosaurs, implies that they represent both countable and uncountable quantity by means of mental magnitudes. ... The system of arithmetical reasoning with these mental magnitudes is closed under the basic operations of arithmetic, that is, mental magnitudes may be mentally added, subtracted, multiplied, and divided without restriction." By "mental magnitudes", Gallistel et al mean what I mean by "analog representations".

### 3.2 *Q&I Modules*

Humans not only employ analog representations of length—they also manipulate them mathematically. For example, we can judge that one length is the sum of two others. This is the kind of mathematical operation that normally requires a cardinal measure. We might say that human analog representations of lengths constitute a kind of nonnumerical cardinal measure, because associated with those analog representations are mental operations that correspond to addition.

Elsewhere (Pollock 1989, 1995) I introduced the notion of a *Q&I ("quick and inflexible") module*. Q&I modules are fast special-purpose cognitive modules that supplement reasoning. For example, if someone tosses an apple to you, you can reach out and catch it. To do this you must predict its trajectory. This is something we are able to do quickly and reliably. If we had to predict the path of the apple by reasoning about parabolic trajectories, it would long since have passed us by before we would be in a position to try and catch it. Q&I modules are fast, but they are also inflexible in the sense that they achieve their speed in part by making assumptions about the environment. For example, our trajectory module works on the assumption that the path of the apple will be unobstructed. If the apple is going to bounce off a piece of furniture, we must wait until we see its new path before we can predict its trajectory.

It is significant that the prediction of trajectories employs analog representations of speed and direction. We could solve the same prediction problem by explicit reasoning (albeit much more slowly), but first we would have to have numerical measurements of speed and direction. We do not usually have such numerical information at our disposal.

Human cognition employs a number of different Q&I modules. For example, most of our reasoning about probabilities seems to be based upon analog representations of probabilities and their manipulation by Q&I modules. It is rare for people to have actual numerical values for the probabilities they cognize about in everyday life. On the other hand, we *can* measure probabilities numerically. One of the burdens of a theory of probability is to tell us how to do that. For example, subjective probability theory attempts to construct numerical measures by comparing our analog representations of probabilities with analog representations of probabilities in games of chance for which numerical measures are readily available. This is similar to the way in which we construct numerical measures of lengths by comparison with rulers.

### *3.3 Numbers*

It seems fairly clear that our normal representation of quantities is analog rather than numerical, and our normal way of manipulating analog representations is by employing various kinds of Q&I modules. The numerical representation of the cardinality of a set may be a built-in mode of representation in the human cognitive architecture, but its extension to the numerical representation of other kinds of quantities appears to be a human discovery or invention rather than a built-in mode of representation. We learn how to assign numbers to other kinds of quantities by discovering how to compare the quantities with standardized quantities that have numbers associated with them in some natural way. Our manipulation of analog quantities then corresponds to the mathematical manipulation of numbers. Some Q&I modules correspond to very sophisticated mathematical computations. For instance, humans can reason about areas in much the same way they reason about lengths. Suppose, for example, that I draw three irregular figures on the board and ask you to draw a figure whose area is the average of the areas of the given figures. The result might be something like figure 2.1. It is extremely interesting that this is a task we can actually perform, and without great difficulty. The results are unlikely to be exactly right, but they will be approximately right (Ariely 2000; Ariely and Burbeck 1995; Chong & Triesman 2003). To perform this same task by measuring the areas and constructing a new figure with the average area of the three given figures would be extraordinarily difficult.

Gallistel et al. (2000) survey the psychological literature and conclude, "There is considerable experimental literature demonstrating that laboratory animals reason arithmetically with real numbers (*that is, analog representations, J.P.*). They add, subtract, divide, and order subjective durations and subjective numerosities; they divide subjective numerosities by subjective durations to obtain subjective rates of reward; and they multiply subjective rates of reward

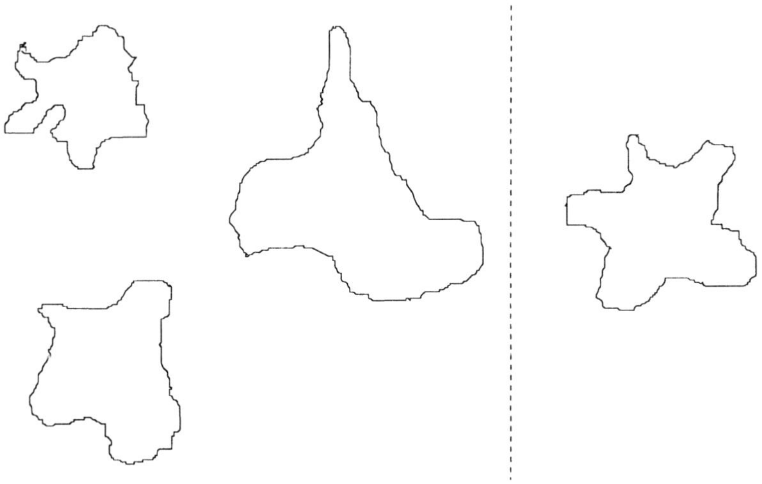

Figure 2.1. Averaging areas

by the subjective magnitudes of the rewards to obtain subjective incomes." The latter amounts to computing analog representations of expected values. Apparently even pigeons and rats can do this.

### *3.4 Decision-Theoretic Reasoning*

Having surveyed some other uses of analog representations of quantities and their manipulation by Q&I modules, it is a small step to the conclusion that human evaluative cognition employs similar tools. Humans can introspect not only that they like one thing better than another, but also that they like it a lot better, or perhaps just a little better. Such comparisons make no sense if we suppose that humans are capable of only ordinal comparisons. It seems undeniable that when we make such judgments we are employing analog representations of value-theoretic quantities.

If human evaluative cognition is to be an implementation of decision-theoretic reasoning, then humans must be able to compute expected values. Expected values are the sums of products of probabilities and values. The computations required for actual cases can seem formidable. However, it is no more formidable than the mathematics required for averaging the areas of irregular figures, and humans can do that quickly and efficiently. I suggest that we are likewise equipped with Q&I modules that enable us to compute expected values with equal ease (although again, with less than total precision). How might we compute an analog representation of the expected value of an action? Here is one way. First we consider the different features that might result from executing it. For each of these features, we *imagine it*, ask ourselves how much we would like or dislike it, introspect the result, and take that as our analog representation of its value. Then we consider how probable each feature is (using analog representations of probabilities), and finally we "mush them all together" to produce an analog representation of the weighted average which is the expected value. The operation of "mushing them all together" is a Q&I module whose function is precisely that of computing expected values.

I find that many philosophers and economists regard this as a radical suggestion. But there is psychological evidence concerning vertebrates ranging from fish to humans indicating that they do compute subjective (analog) representations of expected values (Catania 1963; Harper 1982; Keller and Gollub 1977; Leon and Gallistel 1998). See Gallistel et al. (2000) for a summary of that data.

The upshot of this is that humans really can perform the cognitive tasks required for decision-theoretic reasoning. What they *cannot* do easily is convert the tasks into explicit mathematics and solve the mathematical problems, but that is not necessary. Q&I modules operating on analog representations solve the same problems, just as they do almost anywhere that humans deal with quantities.

In talking about values, it is customary to measure them in terms of "utiles". This can seem puzzling, because no one has ever proposed a numerical scale of utiles or explained how to actually measure values in terms of utiles. This becomes less puzzling when we realize that humans employ analog representations of value in evaluative cognition. No scale problem arises for analog representations. However, the manipulation of analog representations of quantities is generally rather imprecise. For other kinds of quantities having cardinal measures, like length or area, we increase the precision of our reasoning by discovering ways of using numbers in place of analog representations. That is done by devising ways of measuring quantities by comparing them with standard units of the quantity. For example, we measure lengths using rulers. It is important to realize that this bootstraps on a prior ability to manipulate analog representations of lengths. We cannot use a ruler for measuring lengths unless we know that its length does not change when we move it around. You cannot make rulers out of silly putty. And you cannot determine that your rulers are rigid by measuring them with rulers. That leads to an infinite regress. So the ability to judge lengths using rulers presupposes a prior ability to judge lengths non-numerically.

We could likewise improve the precision of "scientific" evaluative cognition if we could discover a way of measuring values numerically. Just as in the case of lengths, this would bootstrap on a prior ability to employ analog representations of value. The use of currency provides the standard mechanism for doing this. Exchangeable currency allows us to construct numerical comparisons of the values of disparate features. This is not a perfect mechanism because the standard way of implementing it results in money changing value over time (inflation), and additivity is not perfect because of the diminishing marginal utility of large quantities of money. But this provides a starting point for the scientific measure of value, and allows humans to convert their decision-theoretic analog cognition into more precise numerical calculations. More needs to be said about this, but I will not pursue it further here.

### *3.5 Theorizing about Practical Cognition*

Apparently humans do not usually engage in explicit decision-theoretic reasoning if that is understood as manipulating numerical measures of values and probabilities. However, it seems likely that human practical cognition uses analog representations to approximate some kind of decision-theoretic reasoning. Furthermore, if we can measure values numerically using a mechanism like currency, then the decision-theoretic reasoning can be made mathematically precise. Thus we can illuminate the logical structure of human practical cognition by considering how such cognition *can* be performed using real-valued measures of utility and probability. Artificial rational agents will not be subject to the representational limitations of human beings in this respect. They can easily be built to introspect numerical values and use them in explicit calculations. And humans can convert their analog reasoning into mathematically precise reasoning if they can produce numerical measures of their values. Accordingly, the bulk of this book will be devoted to an investigation of rational cognition on the idealizing assumption that the cognitive agent is employing numerical cardinal measures of values and probabilities. I will try to note those places where this may make a difference to the theory.

## 4. Conclusions

I have argued on purely computational grounds that binary preferences cannot constitute the primitive evaluative data structure employed by evaluative cognition. The only way to make evaluative cognition work is to have a database of "computationally primitive values" from which other values not included in the database can be computed. I refer to this computation as "the database calculation". This is the only way to avoid having to store more values than it is possible to store. For such computations to make sense, the computationally primitive values must be stored in such a way that it makes sense to perform mathematical computations on the values. The obvious way to do this is to store them in the form of a real-valued cardinal measure. However, for rational decision making, agents must have introspective access to the stored values, and humans cannot introspect real-valued measures of values. This problem is resolved by noticing that humans generally employ analog representations of quantities in place of real-valued measures. The analog representations serve the same purposes, although somewhat less accurately. When we are able to find a way of constructing a real-valued measure, we can use that in place of the analog representations.

Two obvious questions remain. Where do the values stored in the evaluative database come from, and how does the database calculation work?

The first of these questions is taken up in chapter 3, where it is argued that values are discovered by employing a cognitive process I call "evaluative induction". Chapter 5 investigates the database calculation.

# 3
# Evaluative Induction

## 1. The Need for Evaluative Induction

The evaluative database provides efficient storage of values by explicitly storing as little as possible and then computing values for more complex features (situation types) from the stored values. The exact nature of this computation, which I will call *the database calculation,* will be taken up in chapter 4. The entries in the evaluative database represent *computationally primitive values,* in the sense that they cannot be derived via the database calculation from other values. Where do the computationally primitive values come from?

One possibility is that the evaluative database is a primitive constituent of the agent's cognitive architecture—the elements of the database are not derived from anything else. If our interest is in building artificial rational agents, it seems we could stop here. This account tells us how evaluative cognition in such agents can work. We just assign some computationally primitive values and give the agent the ability to compute other values from them.

It takes little reflection to realize that humans do not work this way. We acquire new feature likings and modify old ones on the basis of experience. No one was born liking foreign travel or democratic societies. There probably are some innate feature likings. For example, we naturally dislike being hungry or in pain. But most are learned. I will refer to the mechanism by which new feature likings are acquired as *evaluative induction,*[1] and I will argue that it has a structure that makes instances of it subject to rational evaluation. Accordingly, it can be viewed as a kind of inference.

The suggestion is that the evaluative database is constructed incrementally as the agent acquires knowledge of the world. The incremental construction cannot be random—it must be based upon the agent's discoveries about the way the world works. In other words, there must be a kind of cognition whose purpose is to add elements to the evaluative database in response to acquiring knowledge of the world. This cognition can be viewed as a kind of inference, in which case its operation has the function of deriving the values encoded in the evaluative database from something else. This is what I am calling *evaluative induction.*

---

[1] This is similar to what Millgram (1997) calls "practical induction".

How does evaluative induction work? This can be divided into three separate questions. First, the product of evaluative induction is the values stored in the evaluative database, but just what are these values? They are values the agent attaches to features (types of situation), but can we identify them with some more familiar conative state in human beings? For instance, is the evaluative database a record of the strengths of an agent's desires? I will argue in section 2 that it is not, but that it can be regarded as a measure of the strength of another conative state, which I call "feature liking". Second, what is the information on which evaluative induction operates to produce the values stored in the evaluative database? I will argue in section 3 that evaluative induction appeals to the time course of what I call "state liking". Third, how does evaluative induction use that information to arrive at the values to be stored? I will suggest that this is partly a matter of inductive reasoning and partly a matter of conditioning.

## 2. Human Conative States

I have given a theoretical argument to the effect that humans (and other rational agents) must base practical cognition on a stored database of cardinal values, and these values must be derived from something else by what I have called "evaluative induction". To what does this database correspond in our mental lives, and what are its values derived from? Philosophers are fond of talking about "belief/desire" psychology, and often seem to think that the only conative state in humans is that of desire. However, it does not take much reflection to see that this is false. We can distinguish between at least three different kinds of human conative states. Desires are not the most basic, and they do not play any direct role in the computation of the expected value of an action. I will argue that the denizens of the evaluative database are what I will term "feature likings", and they are derived from "state likings" by evaluative induction.

### *2.1 Feature Likings and Desires*

There are many things that I like. For example, I like eating Greek food. What we like or dislike are features of our state. I will refer to this conative attitude as *feature liking*. Technically, feature liking is a propositional attitude. That is, it is an intentional state whose object is the feature liked or disliked. The intentional state includes a representation of the feature that is its object. I suggest that it is feature likings that are stored in the evaluative database. I will discuss this proposal at length, but first let me contrast it with another proposal that one often hears. This is the suggestion that what is stored in the evaluative database is desires.

Although philosophers use the term "desire" a lot, I am not sure they are really talking about desires. I think in many cases they mean by "desire" what I mean by "feature liking". Taken literally, that is to misuse the term "desire". To see the difference between desires and feature likings, let us begin by considering the functional role of desires in human cognition.

Human beings form desires for various features and then try to achieve them. That is just to say that their desires encode goals. They then engage in means/end reasoning to try to achieve their goals. So I think it is at least roughly true that desires are for things for which you have feature likings and which you adopt as goals.

There are many things I would like to have that I do not adopt as goals. For example, I would like to visit Mars in a flying saucer. I think that would be great fun. But I do not adopt it as a goal because I do not believe there is any reasonable way I could accomplish it. I also like many things that I already have, and I do not adopt them as goals either, because I already have them. For instance, I like having a good mountain bike, but I do not desire it because I already have it.

So desires and feature likings are not the same thing. We might say that to have a feature liking for something is to find it *desirable*, but you do not desire everything that you find desirable. The difference is that desires are for liked features that we have decided to actively pursue. Many liked features are not adopted as goals because we either do not see any practical way to achieve them, or we already have them.

This said, I suspect that many philosophers will agree that what they mean by "desires" is what I am calling "feature likings". So let us discuss feature likings. My proposal is that it is feature likings that are stored in the evaluative database, and they are used both for selecting goals and for evaluating plans in terms of the goals achieved, the costs incurred, and the value of fortuitous side effects.

We can regard the argument of section 1 as an argument that feature likings cannot represent the starting point for evaluative cognition, because there is no way they can all be innate. Most must be derived by evaluative induction from something else. This, however, contrasts with a familiar doctrine often associated with David Hume. Following Hume, philosophers often suppose that desires (a.k.a. feature likings) form the ultimate foundation for practical cognition. Practical cognition aims at achieving goals, which are encoded as desires, but other feature likings are equally relevant to the evaluation of a plan. If a plan fortuitously achieves some desirable result that we had not adopted as a goal, that contributes to its value. The Humean view is that feature likings are not subject to rational criticism. Practical cognition starts with the agent's desires and feature likings, and there is nothing deeper that can be used in their evaluation. All other practical cognition is supposed to be based on this ground level of desires and feature likings. This is the *Humean theory of desire*. But if this is right, how can feature likings be learned?

The Humean theory seems to me to be clearly wrong. Desires can be based on false expectations regarding the feature desired and we regard that as a reason for changing the desire. For example, a woman might think that she would really enjoy foreign travel, and on that basis form the strong desire to engage in it. But when she achieves her goal, she might discover that it is not at all the way she imagined—she is bored by the tedious airplane trips, she hates the unfamiliar food, and she is frightened by being surrounded by strangers speaking a foreign language. The desire for foreign travel quickly evaporates when she discovers what foreign travel is really like. What is more, given her experience, *we would regard her as irrational* if she did not give up her desire. This strongly suggests that desires do not represent the starting point for evaluative cognition.

The same point can be made about feature likings. Consider our foreign traveler again. She begins with a strong desire to engage in foreign travel, and eventually she finds herself in a financial position to achieve her goal. For some years she becomes a globe trotter, traveling to numerous foreign destinations. In fact, she never enjoys her travels, but it may take some time for her to realize that, and in the meantime if you ask her, "Do you like foreign travel?" she will reply, "Oh yes, I really like it". What are we to make of this answer? I think there is a sense in which she is right—she does like foreign travel—and another sense in which she is wrong. The sense in which she is right is that she is in the intentional state of liking that feature. This is just a remark about her current psychological state. The sense in which she is wrong is that she does not have a disposition to enjoy herself when she engages in foreign travel. Although the latter is true, she does not notice that it is and so retains the feature liking. To say that she enjoys foreign travel in the dispositional sense is to assert a statistical generalization about her. It is to say that engaging in foreign travel *tends to cause her* to enjoy herself. One can be ignorant of this sort of causal generalization about oneself.

The intentional state of feature liking is functionally similar to *believing* that the feature that is the object of the liking will tend to be conducive to liking your situation, and it is rationally criticizable by appeal to the same considerations that would make the belief rationally criticizable. But feature liking is a different psychological state from belief, because you can be in either state without being in the other.

If you like a feature, in the sense of having a feature liking for it, but you know that being in a state exemplifying that feature invariably makes you unhappy, then should you regard that feature as having positive value? It does not seem so. It seems that you should regard it as having negative value and regard your feature liking as in some sense "in error". This seems to indicate that feature likings are not the starting point for evaluative cognition either.

It is the distinction between the intentional and dispositional senses of liking a feature that makes feature likings and desires rationally criticizable. Our foreign traveler had a feature liking for foreign travel, and a desire to engage in it, but she was wrong in thinking she would like doing it. This highlights a kind of human conative state that is distinct from both desire and feature liking—*liking* being in a state exemplifying the desired feature. The distinction is an important one, because it seems that desires or feature likings are rationally criticizable by appeal to such *state likings*. If we desire something but know that we wouldn't like it if we got it, we regard that as a criticism of the desire.

In ethics, most desire theories follow Sidgewick in being *informed* desire theories. According to such theories, value attaches to what one *would* desire if one were fully informed about all relevant matters. It is worth noting that the preceding problem can arise even for fully informed desires. Compulsions are normally desires one should not have. Knowing that one should not have the desire to smoke, or eat rich foods, is not sufficient to make these desires go away. The smell of a garlic-laden dish may create an overwhelming desire to partake of it even though one knows that it won't taste as good as imagined and the ensuing gastrointestinal distress will make one regret one's gustatory indiscretion. Conversely, one can know that one would like something if it happened, and have the ability to achieve it, but fail to desire it. For example, I know that I always enjoy skiing when I go, but for some reason that knowledge does not generate a desire to go skiing. It ought to, and its failure to do so is a mark of irrationality. Thus what one *should* (rationally) like or desire need not be the same as what one would like or desire if one were fully informed.

### 2.2 *State Likings*

When our foreign traveler engages in foreign travel but fails to enjoy it, she dislikes her current state. In human beings, such *state liking* seems to provide the court of last appeal in evaluative cognition. If we either like or desire a feature, but being in a state exemplifying that feature does not contribute causally to our liking the state, then it seems that our likes and desires are criticizable. We should not regard being in a state exemplifying that feature as contributing value to our current state. A rational agent who realizes that a feature liking or desire is not conducive to state liking should not pursue it. Conversely, if the agent realizes that a feature is conducive to state liking, then he should form a liking for it.

What kind of a cognitive state is state liking? I described it as "liking one's current state". That suggests that it is an intentional state whose object is the state of the world at a time—a *world-state*. However, I think this is a misleading way of describing state liking. Intentional states are "about" their objects by virtue of containing mental representations of them. For example, if I fear a lion, there is some way in which I am thinking of the lion. When I have a state liking, it does not seem to me that I need be thinking about my current state at all. This suggests that state liking is not really an intentional state. It is better described as a feeling. It is a psychological state characterized by a parameter (the degree of state liking) that varies over time. In this sense it is like numerous other psychological states, including happiness, depression, fear, and so on. We can be happy or depressed *about* something, and we can fear some particular thing, but we can also *feel* happy or depressed or afraid without those attaching mentally to any particular object. Thus although there is a sense of happiness, depression, and fear in which they are intentional states, there is also a sense in which they are nonintensional states. In the latter sense, they are *feelings*. In the same sense, state liking is a feeling. We might put this by saying that state liking is *caused by* our current state, but is not *about* it.

My suggestion is that state liking constitutes the final court of appeal in evaluative cognition. Other kinds of conative states are ultimately criticizable by appeal to state liking. I have said that state liking is a nonintensional state—a feeling, if you like. But just what psychological state is this? Bentham talked about feeling happy, and in discussing our foreign traveler I talked about her *enjoying* foreign travel. I also talked about feeling *satisfied* with one's situation. Can state liking be identified with any of these familiar psychological states? Probably not. State liking is more like a conglomerate of all the different kinds of positive and negative feelings we can have. It is a kind of generic "positive or negative attitude". Antoine Bechara (2003) suggests that, neurologically, it reflects overall activation of the somatosensory cortex.[2] We are familiar with state liking through introspection, but most likely the only way to give a more precise description of it is functionally—by giving a general description of how the psychological state is used in cognition. The sense in which state liking is a positive or negative attitude is that it provides the assessments of intrinsic value used in evaluating actions. So what my claim really amounts to at this point is that (1) there is a nonintensional psychological state that plays the functional role of grounding assessments of intrinsic value, and (2) humans have introspective access to this state. By the latter I do not mean that we have infallible access, any more than we have infallible access to other psychological states like beliefs and pains. There is experimental evidence indicating that there are large individual differences between people's abilities to introspect their conative states (Lane 2000). Interestingly, there are robust gender differences. In general, women are better at it than men.

Every now and then I encounter a philosopher who professes not to know what I am talking about when I talk about state likings. This puzzles me. I am unsure what to make of this disclaimer. Surely all of these philosophers can *sometimes* tell by introspection that they are happy or unhappy, enjoying themselves or feeling miserable, and these are just extremes of state liking. Consider an example. Imagine yourself hiking in the mountains. When you begin you feel strong and vigorous. Your spirits are high. At this point you have an elevated state liking. As you hike along you see wonderful vistas. They please you, and each time this happens your state liking takes a little jump. But then unexpected clouds roll in, and you begin to worry about being caught on the mountain in a storm. As you worry your mood darkens, and your state liking is depressed. You decide to take a shortcut back to your car to beat the rain, so you cut across country and enter the deep woods. But the rain catches you anyway, and you are soon cold, wet, and miserable. Your state liking plummets. Things go from bad to worse as you slip on the muddy ground and find yourself lying full length in the mud. You extract yourself and press on, but your muscles are beginning to ache. You fear hypothermia, and you are beginning to suspect that you are lost. It is almost sunset. Your state liking dips to a new low. But then

---

[2] In cognitive neuroscience this has come to be associated with what is known as "the somatic marker hypothesis", due originally to Domasio (1994).

abruptly you break out of the trees, and at the same time the rain stops. There, stretching out before you, is a panorama of verdant fields covered with wild flowers and framed by one of the most spectacular sunsets you have ever seen. And best of all, there is your car 100 feet away down a well-worn path. Your spirits soar. You have a highly positive state liking, only slightly tempered by the fact that you are still cold, wet, and tired. You think to yourself, "Good hike."

Now I ask you, is there anything you didn't understand about that story? If not, then you understand what state liking is. If you insist that you could see that the sunset was red, but could not introspect that it pleased you and made you feel better, then I begin to suspect that you are "conatively blind", in the sense of lying at the lower tail of the bell curve describing people's ability to introspect their conative states. As remarked above, Lane (2000) observes large variations between subjects. The inability to introspect one's conative states is a recognized clinical disorder known in the psychiatric literature as *alexithemia* (Taylor et al., 1991). Most likely no one is completely conatively blind unless they have suffered damage to the prefrontal cortex, but some are much more aware of their conative states than others. If you really are conatively blind then you are like a color-blind person arguing that there is no such color as green because you can't see it. How can we satisfy the color-blind person? Presumably by pointing out that most people tend to agree about what things are green, so there must be something objective underlying their report of seeing green things. The same strategy can be applied to state liking. Lane musters a great deal of evidence to show that people's introspective judgments of conative states correlate strongly with various kinds of brain activity as measured by fMRI and PET studies. In particular he cites Tranel (1997) as giving evidence that the amygdala is preferentially activated in association with aversive stimuli (negative state liking) and he cites Koob and Goeders (1989) as giving evidence that the ventral striatum is preferentially activated by appetitive and reward stimuli (positive state liking). It cannot be credibly denied that introspective reports about conative states reflect real neurological phenomena.

## 3. Evaluative Induction

I have argued that it is feature likings that are stored in the evaluative database, and that feature likings are criticizable by reference to state likings. I have also argued that feature likings must be derived from something else, by a form of cognition I called "evaluative induction". The obvious hypothesis is that evaluative induction derives feature likings from state likings. An agent is criticizable for liking a feature that is not conducive to state liking or for not liking a feature that is conducive to state liking. This strongly suggests that a feature liking stored in the evaluative database purports to be a record of the degree to which the feature is conducive to state liking. Let's see if we can make this precise.

### *3.1 Cumulative State Liking*

What does it mean for a feature to be "conducive to state liking"? Most features do not occur in an instant of time. For instance, I may attach value to eating a bowl of ice cream. This occurs over a period of time rather than at an instant. So its value cannot be identified with the expected value of the feature liking at any single instant. Furthermore, the occurrence of the feature can have ramifications extending indefinitely far into the future, and these also contribute to its value. Eating a bowl of ice cream may result in an initial positive state liking, followed by a negative contribution to state liking when, because I am lactose intolerant, it gives me indigestion. To compute the value of the feature we must do something like accumulating (integrating) the state liking over time. This means that instantaneous state liking is not treated as value, but rather as rate of value production. The result of accumulating the state liking over time is *cumulative state liking*.

It is obviously difficult to determine what part of an agent's state liking at a time is attributable to any particular cause, but this seems no more difficult than causal attributions in other contexts. We rely upon generalizations confirmed by comparing otherwise similar cases in which a potential cause is present or absent. Still, this is not going to be an easy task, and that will become important below when we turn to how evaluative induction works.

A feature may contribute different amounts of cumulative state liking in different states. We are interested in the degree to which a feature *tends* to do this. Let us define the *abstract value* of the feature to be the average amount of cumulative state liking caused by being in a state exemplifying that feature. This can be cashed out in terms of mathematical expectations.

Making the preceding precise, we want to associate with a feature *S* the amount of cumulative state liking caused by being in *S*. We suppose that in a possible world *W*, for each time *t* there is an amount of state liking at *t* that is caused by *S*. Then the *cumulative state liking caused by S in W* is the result of accumulating that state liking over time.[3] Formally:

$$\text{cumulative-state-liking}(S \text{ in } W) = \int_{-\infty}^{\infty} \text{state-liking-caused-by}(S \text{ in } W \text{ at } t)\, dt$$

The *abstract value* of the feature is then the mathematical expectation of the amount of cumulative state liking caused by being in a state exemplifying that feature. Formally, where *W* ranges over possible worlds:

abstract-value(*S*)

[3] Should future values be discounted? I see no reason to think so (other than limited life expectancy, which is already factored in when a probability is assigned to whether the feature will occur), but there is no theoretical obstacle to doing so.

$$= \text{EXP}\big(\text{cumulative-state-liking}(S \text{ in } W) \,/\, W \text{ is a possible world}\big).$$

Recall that EXP was defined in chapter 1. The mathematical expectation of a quantity is a weighted average of its possible values, weighting each by the probability of its being the actual value. As remarked there, whenever we use the EXP notation, we must make clear what kind of probability is being employed. Here my intention is to use nomic probability, which will be discussed in chapter 7.

### 3.2 *Correcting State Liking*

A complication arises from the fact that our state liking is influenced by our beliefs. If we have false beliefs, or lack some true beliefs, we may like our state better or worse than we would if our beliefs were corrected. If we like being in states of a certain type (exemplifying a certain feature) only because of false beliefs we would have in such states, should we regard that as making the feature valuable? For example, Nozick (1974) discusses the "experience machine". This is a machine that can make the world seem any way the agent chooses, and it can be configured so that the agent does not know he is in the experience machine once it begins operation. Suppose I have some concrete goal, for example, to construct a unified field theory for physics. In fact, the goal is beyond my ability. However, by subjecting myself to the experience machine I can make myself *think* that I have accomplished this goal, and that will make me quite content. Is this a reason for subjecting myself to the experience machine? Many seem to think it is obvious that it is not, on the grounds that what I want is to *really* construct a unified field theory—not just *think* I have.

But I am not sure that the preceding is correct. Consider a different example. Suppose I have had some traumatic experience that I cannot stop thinking about, and it is making me miserable. If I am told that I can be hypnotized so that I will forget all about the experience, and this will alleviate my misery, would it be irrational for me to do that? It seems to me not. This is a case in which it is better not to know. Or consider an exploratory hike in new territory. Not knowing what is around the next bend can make the hike more exciting.

In light of these conflicting examples, it is unclear to me whether, in evaluating a feature, we should consider the agent's current state liking, or consider counterfactually what the agent's state liking would be if he were fully and accurately informed about his state. I am inclined to think we should do the former, but if we decide we should do the latter, that can be accommodated in the definition of *cumulative state liking caused by S in W* by understanding *state liking caused by S in W at t* as the state liking the agent would have if he were accurately informed about the relevant aspects of his state. However, I am going to leave open whether such an understanding is required for a correct theory of rational decision making.

### 3.3 *The Evaluative Database*

The evaluative database is a database of feature likings. Feature likings are rationally correctable by reference to abstract values. The abstract value of a feature $S$ is the degree to which $S$ is conducive to state liking. If we have a feature liking for $S$ but discover that it does not reflect the degree to which $S$ is conducive to state liking, then our feature liking is criticizable and we should employ the believed abstract value in decision-theoretic deliberations rather than the stored feature liking. If possible, we should correct the entry in the evaluative database. This suggests that the feature likings stored in the evaluative database are just measures of abstract value.

But there is a problem with this suggestion. Mathematically, the definition of abstract value makes perfectly good sense. But the mathematical definition is of little help in actually computing the value. The problem is three-fold. First, in a given situation, it is going to be very difficult to determine at any particular time how much of the state liking is attributable to a feature $S$. Second, even if an agent can solve the first problem, how does he know how to accumulate the state liking over time, that is, how does he do the integration involved in the definition of "cumulative-state-liking"? Third, even if the agent can somehow compute the cumulative state liking caused by $S$ in different possible worlds, how does he average them over possible worlds (that is, how does he do the integration in the mathematical expectation involved in the definition of "abstract value")? The latter certainly cannot be done by surveying all possible worlds. Computing abstract values in accordance with the definitions looks like a hopeless task.

However, we face and solve similar problems elsewhere. For example, the Humane Society advertises that an un-neutered tomcat will, on the average, father 6.4 kittens in one year. This is a claim about mathematical expectations. How do they know that it is accurate? First, they have the problem of knowing that a particular tomcat has fathered a particular kitten. Then they have the problem of knowing how many kittens it fathers in a year. Then they have the problem of averaging that figure over all tomcats. So where does their number come from? The answer is that they do not have to do all these calculations. Standard statistical reasoning allows them to estimate the total number of "unplanned" kittens each year, and the total number of un-neutered tomcats, and they divide the first number by the second.

Can we do something similar to compute abstract values? We can, but it will not be simple. First, we could establish inductively that features rarely affect state liking outside of a certain time interval surrounding the occurrence of the feature. For example, eating a bowl of ice cream may affect state liking for a few minutes, or even for a few hours, but it will not usually have ramifications over longer periods of time. We can discover this by comparing state likings at times beyond the end of the interval. We examine a number of cases in which the feature occurs and a number of cases in which it does not, and see that on the average there is no difference in state liking after the end of the interval. Next we can monitor the state liking within the interval, comparing cases in which the feature occurs and cases in which it does not. This will give us an average figure, which we can take to be an estimate of the abstract value. This sort of investigation could be carried out in a scientific way by a psychologist. It would be complex because care would have to be taken to factor out confounds, but it is in principle possible.

Obviously, humans do not ordinarily compute abstract values the way the scientist would. I observed in chapter 2 that explicit reasoning is slow, so epistemic cognition supplements it with Q&I (quick and inflexible) modules. These are special purpose modules that are able to process certain kinds of information and draw certain kinds of conclusions rapidly, typically by making assumptions that are usually, but not always, satisfied. We employ many different kinds of Q&I modules. These include modules for predicting trajectories, inductive and probabilistic reasoning, various kinds of spatial reasoning, and perhaps much more. Among our Q&I modules is one for what might be called "intuitive induction". It appears that most statistical generalizations believed by humans are the result of a Q&I module that summarizes the data as it is collected, without compiling an exhaustive record of it for later perusal. The resulting statistical generalizations may not be completely accurate, but they are often all we have.

So one suggestion is that the feature likings stored in the evaluative database are the results of such intuitive induction. However, intuitive induction produces *beliefs* (with analog representations of probabilities or expectations, or in this case, abstract values). In humans, the residents of the evaluative database are likings, not beliefs. They are conative responses to features rather than beliefs about features. They are correctable by reference to beliefs about abstract values, but they are not themselves beliefs. If they were, we would not have irrational feature likings wherein our liking is not commensurate with what we believe the abstract value to be.

Although feature likings are not beliefs, many feature likings do seem to be rather direct expressions of beliefs about abstract values. For example, when I am teaching an undergraduate course I like to give an exam fairly early in the semester because I have learned that doing so makes the students more diligent. Similarly, I like to get the oil changed in my car every 5,000 miles, because I have learned that the car will wear out more slowly. These are genuinely things I like to do (although not things I like doing). The liking is motivated by a belief about abstract values, but it is not identical with the belief. If I change my mind and decide it is better not to give an early exam, it is likely that my feature liking will change too. But sometimes there is a time lag. I may find myself scheduling an exam and then realizing that I am "forgetting" that I no longer believe that is a good thing to do.

One way of generating feature likings is by direct inductive reasoning about abstract values. For example, I have discovered that I enjoy teaching certain courses and I dislike teaching others. This is stored in the form of feature likings which I access when deciding what to teach next year. Another way of generating feature likings is by means-end reasoning. I have discovered that giving an early exam makes the students more diligent, and student diligence is something I value (attribute a positive abstract value to), so I can infer that giving an early exam has a similar positive abstract value (attenuated by the probability of its achieving its goal).

Feature likings that are generated by beliefs about abstract values are fairly easily updated when we change our beliefs. But we also exhibit feature likings that are more resistant to updating. I may like eating garlic laden dishes. If I discover this to have adverse gastrointestinal effects, I may conclude that eating such dishes has a negative abstract value. But this does not automatically make the feature liking go away. I still like eating those dishes, and may strongly desire them, but rationality dictates that my belief about abstract values should override my stored feature liking and I should avoid eating the dishes.

What is the difference between the feature likings that are resistant to doxastic updating and those that are not? I suggest that the former are the result of conditioning rather than induction or means-end reasoning. Conditioning is an epistemologically more primitive process than intuitive induction, but it can be regarded as aiming at a similar result. Conditioning attempts to attach a degree of liking to a feature that reflects the average state liking the agent experiences around the time it is in a state exemplifying that feature. This is normally correlated with the abstract value of the feature, but not always. Anomalous statistical correlations can condition agents to like features that are associated with causes of state liking without themselves being causes of state liking.

Conditioning can establish feature likings without the agent engaging in much, if any, explicit epistemic cognition. It is an automatic process that goes on largely in the background. In this sense it is more efficient and more easily accomplished than "more cognitive" processes like either intuitive induction or explicit scientific investigation. On the other hand, it is also more prone to error, in the sense of producing feature likings that are not reflective of abstract values. Its greater efficiency makes its use in an agent desirable. If an agent had to acquire all of its feature likings as a result of epistemic cognition, even employing Q&I modules like intuitive induction, it is doubtful that it would be able to acquire enough of them fast enough to get around in the real world. However, the potential inaccuracy of conditioning also makes it desirable to equip a cognitive agent with means for correcting conditioned feature likings by appeal to reasonable beliefs about abstract values. I take it that this explains both why feature likings in human beings are not simply beliefs about abstract values, and also why we try to correct our feature likings when we believe that they do not accurately reflect abstract values.

### *3.4 Conative Conditioning*

I will refer to the kind of conditioning that establishes feature likings as *conative conditioning*. The nature of conative conditioning is an empirical matter, and should be investigated as such. However, I will make some preliminary and speculative observations. First, conative conditioning does not have the form of classical conditioning, but it is natural to suppose it is ordinary operant conditioning where the conditioned response is a conative response and state liking provides the reinforcement. However, conative conditioning doesn't quite have the form of operant conditioning either. Negative state liking does not just condition the agent not to have a positive conative response to a feature—it conditions the agent to have a negative conative response. This cannot be established by ordinary operant conditioning.

Conative conditioning is further complicated by the fact that it allows "post hoc conditioning". A person may engage in a risky activity and enjoy it at the time, but then afterwards spend some time thinking about "what might have happened". He imagines dire situations in which he would have experienced highly negative state likings, and that has the power to recondition his initially positive feature liking, making it negative. For instance, I have introduced many of my graduate students to mountain biking. Some of them profess to enjoy it when they try it, but when they think about it afterwards and think about all the catastrophes they believe might have befallen them they form an aversive feature liking for the activity and decline to try it again.

It is noteworthy that the strength of this effect is modulated by how likely we think the untoward outcome is. I can always imagine bad outcomes of any activity, but because I regard them as very unlikely they do not incline me to dislike the activity. Going to the store to buy milk could result in an auto accident and my being crippled for life, but that does not make me averse to going to the store to buy milk. This indicates that post hoc conditioning is a way of taking account of contributions to the expected value of an action that occur some time after the action.

Post hoc conditioning may be the most important mechanism we employ for correcting conditioned feature likings to bring them into line with actual abstract values. Conditioning based simply on current state liking tends to be short sighted, because we do not think about the long term consequences until later. In fact, it may not be initially obvious that an activity can have a particularly untoward consequence. A fair amount of cognitive effort may be required for that discovery. Thus it was very difficult to discover that smoking causes cancer. A person may have been conditioned to like smoking because it was pleasurable. When he later learns (or becomes convinced) that smoking causes cancer, he is faced with the problem of changing his feature liking to bring it into line with what he now conceives the abstract value of smoking to be. Post hoc conditioning provides a mechanism he can employ deliberately for this purpose. Whenever he thinks about smoking, he can intentionally contemplate what it would be like to get lung cancer and how much he would dislike that.

### 3.5 *Evaluative Induction*

The evaluative database stores values that purport to reflect the abstract values of features. For purely physiological reasons, some features make predetermined contributions to state liking. These include physiological states like being hungry or being in pain. Thus there is no reason why feature likings for these features cannot be innate, that is, stored in the evaluative database from the inception of the agent. However, other feature likings must be established by evaluative induction. Now we can say what evaluative induction amounts to. It consists first of either a conditioning process that attaches feature liking to a feature or a reasoning process about abstract values. Second, it includes a mechanism for updating stored feature likings in light of explicit beliefs about abstract values. By building an agent in this way, we make it possible for it to acquire "usually approximately correct" feature likings fairly easily, but also to get more accurate feature likings in those cases in which it is able to perform the requisite epistemic cognition.

Ideally, we would want the cognitive architecture of an agent to be constructed so that feature likings are automatically updated in response to explicit beliefs about abstract values. It appears that the human cognitive architecture works imperfectly in this respect. We have difficulty correcting our feature likings by appeal to "coldly epistemic" conclusions. Thus, on the one hand we suffer from compulsions to do things that we know to be counter-productive to our state liking, and on the other hand we may know that an activity like skiing is conducive to our state liking but find that we remain emotionally indifferent to it in the sense that our knowledge does not lead us to acquire a feature liking. It seems desirable to have a cognitive architecture that enables an agent to update its feature likings more efficiently in response to beliefs about abstract values, but in chapter 4 I will suggest that this may by hard to accomplish.

There is an alternative to using knowledge of abstract values to update the degrees of feature likings stored in the evaluative database. The role of feature likings is to provide the values needed for decision-theoretic reasoning. But for that purpose, we need not correct the feature likings stored in the evaluative database. Instead, the rules for rational decision making can ignore the evaluative database when the agent has an explicit belief about abstract values, and only use the values in the evaluative database in those cases in which the agent has no such explicit beliefs. If decision making is done in that way, then the storage and retrieval of information about abstract values will be less efficient than if it were all stored in the same way in a single evaluative database, but the erroneous values in the evaluative database will not lead to incorrect conclusions. Human decision making appears to combine both of these cognitive strategies, but neither one works as well as we might desire. Just as we find it difficult to correct feature likings, we also find it difficult to ignore feature likings when we have conflicting beliefs about abstract values—think again of compulsions. I will suggest a possible explanation for this phenomenon below.

## 4. Evaluative Induction as a Q&I Module

The preceding remarks were largely about human evaluative cognition. Some aspects of this cognition proceed automatically, and others are under our control (at least to some extent). Only the latter fall within the purview of rationality. It is useful to compare evaluative cognition with epistemic cognition. Think, for example, about the role of vision in epistemic cognition. At least in humans, visual input at the optic nerve is passed to a module that produces a visual image. The visual image is the input to those aspects

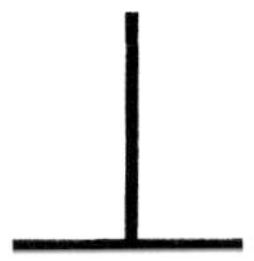

Figure 3.1. The horizontal and vertical lines are the same length

of epistemic cognition that are under the control of the agent and subject to rational evaluation. The agent has a certain amount of control over what inferences he makes on the basis of the image, but the production of the image proceeds entirely automatically (Pylyshyn 1999). Rationality begins where the automatic processes end (Pollock 2007).

It was noted above that many of our beliefs are produced by Q&I modules rather than explicit reasoning. For example, we employ a Q&I module for comparing the lengths of visually perceived lines. Consider this module a bit more carefully. One line can look longer than another, and that inclines us to believe that it is longer. However, there are familiar visual illusions

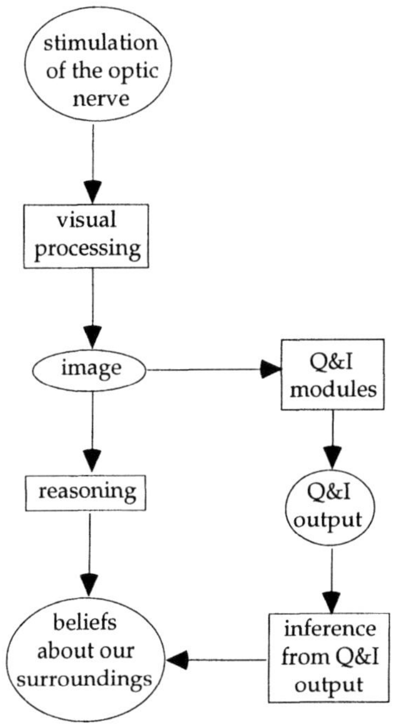


Figure 3.2. Visual cognition

that can produce this effect when the lines are actually the same length. Consider figure 3.1. The vertical line looks longer than the horizontal line, but if we measure the lines carefully we will find that they are the same length. Having made the measurement, we should believe that they are the same length despite the fact that one looks longer. That is, beliefs produced by explicit reasoning should, rationally, take precedence over the output of Q&I modules. Q&I modules are shortcut devices to be used when we do not have time to be more careful. This is sometimes accomplished by having the output of the Q&I module be something other than a belief. In this case, the output is that one line "looks longer". An inference is made from the output of the Q&I module to a belief about the relative lengths of the lines, and that inference can be defeated by contrary information derived from explicit reasoning. Note also that Q&I modules are normally introspectively impenetrable. We cannot monitor their operation—only their output. Functionally, the reason for this is that we have no control over them so there is no reason for our cognitive system to give us the power to monitor them.

So the picture of epistemic cognition that we get is one in which certain automatic processes produce output (images) that is the input to explicit reasoning and Q&I modules. The latter produce beliefs jointly, with the reasoning taking precedence over the Q&I modules when there is a conflict. We might diagram this as in figure 3.2. Rationality only applies to the stages of cognition following the production of the image. The Q&I modules themselves cannot be assessed as to their rationality, but an agent who declines to override a Q&I module when he has good reason for doing so will be judged irrational. For instance, if a person carefully measures the lines in figure 3.1, but still refuses to believe they are the same length, he is

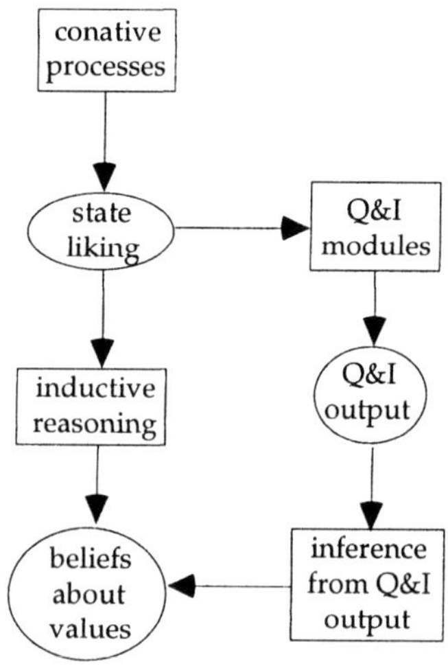


Figure 3.3. A simple model of evaluative cognition

being irrational.

Now let us turn to evaluative cognition. It would be nice to have a similarly tidy distinction between the automatic processes that provide the input to rational cognition and the inferential processes that are subject to rational evaluation. A natural model would begin with a "conative module"that produces state liking. On this model, state liking would be the input to rational cognition, and then explicit reasoning supplemented with Q&I modules would produce the agent's beliefs about values. This is diagrammed in figure 3.3.

Can this model accommodate the fact that in human cognition, what is stored in the evaluative database is likings rather than beliefs, and they are often produced by conditioning rather than inductive reasoning? This can be accommodated if we take beliefs about values rather than the feature likings to be the ultimate product of evaluative cognition. Then the composite process of establishing feature likings by conditioning and using them to infer values can be viewed as a Q&I module. In figure 3.4, this produces a diagram of evaluative cognition with essentially the same structure as figure three. However, in figure 3.4 there are two additional connections. First, general beliefs about values are stored as feature likings and then retrieved as needed by inference from feature likings. Thus we get a loop from beliefs to feature likings and then back to beliefs. Second, when beliefs about values conflict with feature likings that prove resistant to updating, rationality requires us to give preference to the latter, but it also requires us to attempt to "correct" the stored feature likings, perhaps by post hoc conditioning. This is because the evaluative database contains *stored* feature likings, not

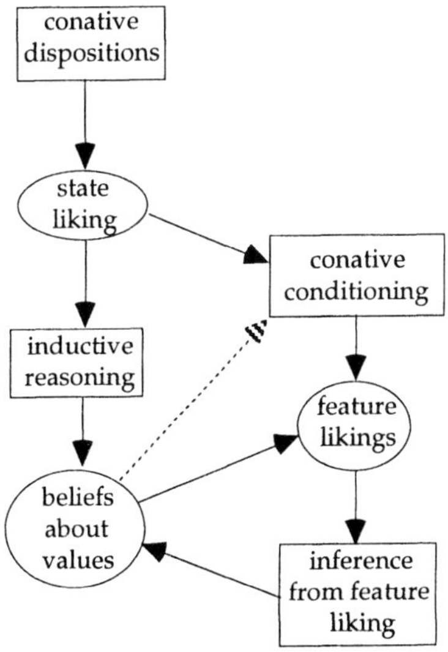


Figure 3.4. Evaluative cognition

just momentary feature likings. This correction process is indicated by the dashed arrow. Viewing the use of feature likings as a Q&I module, this is different from the way Q&I modules are used in visual cognition. In visual cognition, the outputs of the Q&I modules are transient. They are not stored and used repeatedly for drawing new conclusions, so there is no need to try to change the outputs themselves. But in evaluative cognition, the feature likings are stored, so it is desirable to change them rather than just override them.

The rational mandate to try to correct our feature likings reflects the fact that although we can think of conative conditioning as a Q&I module whose output is feature likings, we have more control over conative conditioning than we do over most Q&I modules. We change feature likings by reconditioning ourselves, and that is something we can try to do deliberately. To this end, we can harness the automatic process of conative conditioning by feeding it tailored inputs aimed at changing the output.

A consequence of this account is that what I am calling "evaluative induction" is not a purely rational process. It has elements that fall within the purview of rationality, but it also has elements that proceed automatically and are not subject to rational evaluation. In particular, the process of conative conditioning is automatic, but rationality governs what we do with the feature likings that it outputs, and rationality governs when we should try to harness conative conditioning to recondition feature likings by feeding it tailored inputs.

## 5. Conclusions

I argued in chapter 2 that evaluative cognition must proceed in terms of a stored database of values. I have now argued that these values should be real-valued measures (or analog representations) of feature likings. Features likings, in turn, derive from the time course of an agent's state liking. Ideally, feature likings should represent the expected value of the cumulative state liking caused by being in the states exemplifying that feature. However, it is computationally impractical to require agents to have rationally justified beliefs about such expected values before they undertake practical cognition. Degrees of liking for some features, like being in pain or being hungry, may be built into the evaluative database innately. But many will result from conative conditioning. We can view conative conditioning as a Q&I module for discovering expected values. More accurate assessments can be obtained by what I called "intuitive induction", and highly accurate assessments can be obtained by employing correct scientific procedures. However, it will be rare that the agent has time to pursue the latter. "Evaluative induction" is a generic term that includes all of these methods of establishing feature likings and beliefs about abstract values.

# 4
# Some Observations about Evaluative Cognition

This chapter consists of some miscellaneous observations about evaluative cognition. They are of some interest, but they are inessential to the rest of the theory, so I have collected them into their own chapter and the reader can skip them if he wishes.

## 1. Liking Activities

We have two kinds of feature likings. We like or dislike things that happen to us, and we like or dislike doing things. Although we talk about liking ice cream or liking democracy, this is short for liking eating ice cream and liking living in a democracy. Reflection suggests that my negative feature likings are more often for things that happen to me, although I do have negative feature likings for activities like grading exams or going to the dentist. Conversely, it seems to me that my positive feature likings are more often for activities than for happenings. I like drinking good coffee, riding my mountain bike, seeing a play, conversing with friends, and so on..

Feature likings for activities play a particularly important role in practical cognition. When we have the opportunity to do something that we like doing, we often choose to do it *because* we like doing it. We treat a positive feature liking for an activity as a reason for engaging in it. Conversely, we treat a negative feature liking for an activity as a reason for avoiding it. Of course, the activity can have consequences that outweigh the positive or negative feature liking for it. For instance, I grade my students' exams even though I dislike doing so. This is because not grading them would have consequences that I regard as worse. But in the absence of contravening considerations, I often choose actions just on the basis of whether I like doing them.

What is the justification for basing decisions on feature likings for activities? Feature likings aim at encoding abstract values, so having a feature liking for an activity should encode the fact that engaging in the activity has a particular abstract value. But the abstract value of an activity is the same as its "generic" expected value, that is, the expected value of engaging in such activity in unspecified circumstances.[1] Accordingly, this can be used as a defeasible indication that the activity is or is not a good one to engage in under the present circumstances. So this is a shortcut for decision-theoretic reasoning. It is quite an important shortcut because we often find ourselves faced with decisions in which we cannot make reasonable estimates of the probabilities of various outcomes, and so cannot explicitly compute expected values. However, we do not have to know the probabilities if we can estimate the expected values directly rather than computing them, and that is what feature likings for activities enable us to do. We can regard this as another Q&I module. It takes feature likings as inputs and produces assessments of expected values. Thus we find feature likings playing an important dual role in evaluative cognition. We employ two different Q&I modules that are based on feature likings. We can diagram the structure of the resulting cognition as in figure 4.1. This embellishes figure 3.3.

## 2. Evaluating the Human Cognitive Architecture

I have been describing various aspects of the human cognitive architecture. As I indicated in chapter 1, the rationality of individual cognitive performances is assessed by reference to the agent's fixed cognitive architecture. From this perspective, we cannot criticize an agent as acting irrationally if what we find objectionable is an aspect of his built-in cognitive architecture over which he has no control. But we can step up a level and evaluate the architecture itself in terms of how well it contributes to the satisfaction of various design goals. Looking at humans, we can ask, "Could we build a better agent?" There are a few respects in which it seems that the human architecture for evaluative cognition might be improved.

One obvious improvement would be to measure quantities using real numbers rather than analog representations, and do the mathematics accurately. This is easily accomplished in artificial agents, but in the case of humans evolution failed us.

There is one glaring respect in which the human architecture seems problematic. If we have an explicit belief about abstract value that conflicts with a feature liking, we should make use of the belief and ignore the feature liking in our decision-theoretic deliberations. This is something that humans often find difficult to do. This seems to be a respect in which we could build a better agent. Here it seems likely that we are suffering from our phylogenetic origins and the fact that conation is based on some of the older parts of the brain.

How about the processes involved in establishing feature likings? Sometimes feature likings are based on beliefs about abstract values, but the

---

[1] As indicated in the next section, feature likings often encode functions that describe how the abstract value depends upon certain parameters. This generates an even more useful estimate of the expected value of the activity.

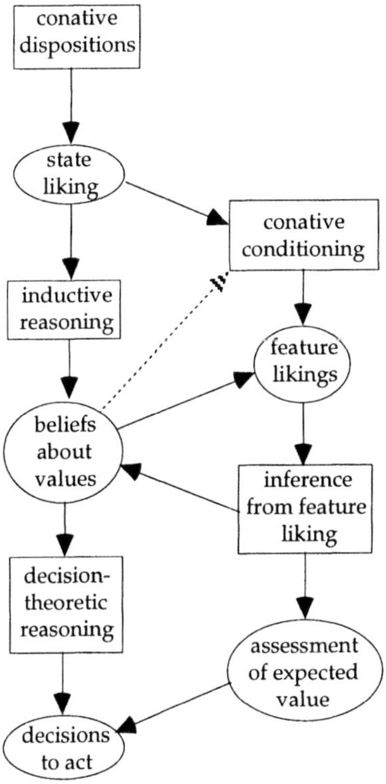


Figure 4.1. Two roles for feature likings

human cognitive architecture also makes heavy use of conative conditioning as a Q&I module. Something like that is probably essential in any agent operating in a complex environment. Ideally, we might prefer that an agent use epistemic cognition to acquire beliefs about abstract values, and then use those beliefs in decision-theoretic reasoning. However, it is doubtful that any agent operating in the real world can perform the requisite epistemic cognition quickly enough to acquire all the beliefs it would need about abstract values. So the agent must be equipped with some kind of shortcut procedure and equipped with cognitive mechanisms for overriding the shortcut procedure when the desired beliefs about abstract values become available.

In the human cognitive architecture, the output of the shortcut procedure is feature likings. These must be stored, because conative conditioning proceeds over an extended period of time and the outputs (feature likings) will evolve slowly in response to new inputs. We access these stored feature likings repeatedly to draw conclusions about values, so they can be viewed as constituting a database. As such, it becomes desirable to update the database when we discover that the stored values disagree with what we believe the actual abstract values of features to be. This is another respect in which the human architecture seems imperfect. We have difficulty updating our feature likings, and are not able to do it just by forming the belief that the feature likings should be different.

Could we build a better agent by having beliefs about values automatically update feature likings? That is more complicated than it might at first seem. The difficulty is that feature likings sometimes have complex structures. A feature liking need not store a constant value. Feature likings are often context sensitive. I like food better when I am hungry, I like exercise better when I am not tired, I like watching movies better when I haven't just watched ten movies, and so on. Most positive feature likings tend to weaken in situations in which they were recently instantiated. This is a way of implementing something like the golden mean—pursue good things in moderation. On the other hand, some negative feature likings (for example, disliking pain, or extreme cold) remain more constant. From the perspective of agent design, this all seems desirable. Biologically rooted negative feature likings tend to attach to things the agent should try to avoid whenever they are in the offing. But on the positive side, agents have a variety of needs. We do not want an agent expending all of its effort pursuing the single thing it likes most while ignoring everything else, and this is accomplished by having recently instantiated feature likings diminish in strength. Presumably, this also reflects the abstract values of the features, which contribute more to state liking when they have not been recently instantiated.

The same effect could be achieved by storing general beliefs about values in an evaluative database, where the values are described as functions of various parameters. Thus the abstract value of food is a function of how hungry we are, and the abstract value of watching a movie is a function of how often we have watched movies recently. The difficulty with this is that the functions are going to be fairly complex. It will be hard to discover them inductively, and even encoding them explicitly may be difficult. Feature likings seem to provide a method for encoding such complex functions using something more like connectionist networks than explicit function representations. Connectionist networks can often be trained to encode very complex functions that we would have great difficulty representing explicitly and great difficulty discovering by inductive reasoning.

If it is true that feature likings rely upon something like connectionist networks to encode complex functions, updating them in response to beliefs about abstract values will tend to be difficult, because those beliefs will likely not contain as much information as the networks. It may be that the only way to update feature likings without losing valuable information is by reconditioning them, which seems to be what humans rely upon. So the human cognitive architecture may be a better solution to the design problems than it might at first appear. The updating problems that we encounter are hard problems that are going to be difficult to solve by any other means. So even if we find it difficult to update feature likings in some cases in which they clearly need it (for example, compulsions), there is no obvious alternative that will work better in these cases without working worse in other cases.

Another aspect of human evaluative cognition that may initially seem puzzling is that explicit beliefs about values are stored as feature likings. It is easy to understand why conative conditioning produces feature likings, but why not store beliefs about values as beliefs rather than converting them into feature likings and then converting them back to beliefs when we engage in decision-theoretic reasoning? I suggest that the answer is that we rarely do retrieve the same values we store. The values needed for decision-theoretic reasoning are typically those of composites of features, and they are obtained by employing the database calculation. The logical details of the database calculation are discussed in chapter 4. But for implementing it, all of the values must be expressed in a common format. It seems likely that the database calculation is implemented in humans by employing a single system, again something like a connectionist network, that handles both the computation of current features likings (sensitive to various parameters) and the computation of the value for the composite of features.

## 3. State Liking

### *3.1 Is State Liking the Only Value?*

I have argued that state liking provides the pillar on which human evaluative cognition is built. We might reasonably say that the goal of practical cognition is to keep the agent's state liking high. This sounds suspiciously like a Benthamite theory of value. Bentham urged that happiness is the only intrinsic good, but he was roundly criticized on the grounds that what we value is not happiness alone but also things like living in a democratic society, interacting with friends, eating ice cream, and so on. Happiness might play a causal role in getting us to value those things, but it is not happiness itself (or not happiness alone) that we value.

This criticism of Bentham is surely right. But the present theory is not subject to a similar objection. To say that the cognitive system aims at keeping the agent's state liking high is quite different from saying that this is the sole aim of the agent. The aims of the agent are to achieve the things it values, like democracy, friendship, and ice cream. The system achieves its end by getting the agent to value these things. Talk of the aims of the system, as opposed to the aims of the agent, uses teleological language to describe the way the system works. Practical cognition aims at keeping state liking high in the same sense that the pumping of the heart aims at moving blood through the body. It is notoriously difficult to say just what this means, but it is clear that it makes some sense and that we often know such observations to be true.

### *3.2 Conative Dispositions*

An agent implementing the kind of evaluative cognition I have been describing must be equipped with a conative mechanism that produces various degrees of state liking. We can think of this as functioning in terms of *conative dispositions* to like one's current state in response to various inputs. Several different conative dispositions may be activated at one time, in response to different inputs, and their outputs are combined to form an overall state liking. What are the inputs to the conative dispositions? Obviously, the human cognitive architecture includes conative dispositions that are responsive to some nonconceptual inputs. Physiological states, like hunger, thirst, fatigue, sexual arousal, pain, feeling hot or cold, and so on, can affect state liking without the agent forming any beliefs. These "physiological" conative dispositions provide an initial source for state liking. It also seems clear that our state liking is affected by our beliefs about the world. One way this is achieved is by having conative dispositions to like my current state when I have certain kinds of beliefs about it. I will refer to these as *doxastic conative dispositions.* In the next section I will discuss other mechanisms for achieving this as well.

Among our doxastic conative dispositions, some might be innate and others learned. However, the innate ones, if there are any, create an implementation problem. Beliefs require concepts. They involve thinking about the world in some particular way, and to think about it we must have a mental representation of what we are attributing to it. Presumably some concepts are innate constituents of our cognitive armory, but most people believe that others are learned. Famously, Fodor (1975) maintained that all concepts are innate, but that has not been a popular view. It is hard to muster a compelling argument about this one way or the other without a theory of concepts, and at this point there is none that I am prepared to endorse. Still, I am strongly inclined to suppose that most concepts are learned, and that they are learned on the basis of experience of the world. What inclines me to believe this is that most concepts do not have literal definitions in terms of simpler concepts. They have a different kind of structure, which seems to give beliefs involving the concept a constitutive role. For example, could you equip an agent, independent of experience, with the concept of democracy? I do not see how. It seems that you learn that concept by learning things about democracy. Of course, "democracy" is an extreme case, but it seems unlikely that more mundane concepts like "mother" or "father" will prove much simpler. For artificial agents, we might provide the requisite knowledge of the world by building an elaborate theory of the world into the agent innately. But there are two problems with that approach. First, building the world theory is a formidable task—probably impractically so. Second, the resulting cognitive agent will be brittle, in the sense that it will be unable to function in a world differing at all from its built-in world theory. Clearly, human beings do not work in this way to any great extent. They acquire most of their world theory through learning, and they acquire the concepts required to think about the world as part of that learning. It seems likely that any general-purpose cognitive agent capable of functioning in unconstrained environments must work similarly.

If an agent has any innate doxastic conative dispositions, they must appeal to beliefs that can be formulated using only concepts that can be built into the agent from the start. Simple concepts might be built in, and there may be conative dispositions attached to them. Some such mechanism is presumably involved in a field mouse's innate fear of a hawk's shadow. However, the restriction to concepts sufficiently simple to be innate imposes severe restrictions on the complexity of the beliefs to which innate conative dispositions can appeal. There is probably no way to build in fear of the hawk itself, but it is possible to build in fear of certain shapes because the concepts of those shapes can be innate rather than learned. Agents designed to work in only narrowly circumscribed environments can have a built-in world theory and the concepts that go with it, and conation in such an agent can make use of a rich array of built-in concepts. But as we remove constraints on what environments the agent should be able to function in, the epistemic cognition of the agent must become more sophisticated so that it can build its own world theory in response to experience. Then the resources available to conation become correspondingly more impoverished because the agent has fewer innate concepts for use in built-in conative dispositions.

The argument against having built-in conative dispositions responsive to complex concepts is more compelling when applied to concepts pertaining to the external world than when applied to concepts pertaining to the cognitive agent itself and to other similar agents. It is undesirable to build too much of an a priori theory of the external world into a cognitive agent, because that precludes its being able to deal with environments that conflict with its built-in theory. However, the same argument cannot be applied to building in an a priori theory of the agent itself. The basic structure of the agent is fixed and will be the same regardless of its external environment, so there is no obvious reason for not equipping the agent with a priori concepts descriptive of itself and other agents like itself. These concepts might be referenced by built-in conative dispositions, and this may play an important role in the sociobiological aspects of human conation. Examples of the latter would include our natural conative responses to property, revenge, or various moral dimensions of our circumstances. In this respect, it is noteworthy that capuchin monkeys and other rather primitive primates share some of these seemingly sophisticated sociobiological dispositions (Brosnan and de Waal 2003; Brosnan, Schiff, and de Waal 2005).

The combination of conative responses to physiological states and conative responses to beliefs involving simple (innate) concepts of the external world and possibly more complex concepts pertaining to ourselves seems like a rather blunt instrument for grounding the evaluative cognition of a rational agent, but it is hard to see what else could be involved.

### 3.3 *The Effects of Feature Likings on State Likings*

The most straightforward design for a cognitive architecture would begin with some built-in conative dispositions, and then add mechanisms for fine-tuning them in the light of experience so that they better serve the architecture's design goals. It is easy to see how built-in conative dispositions of the sort described above should subserve these goals. But it is harder to see how the agent's cognitive performance can be improved by allowing experience to fine-tune state liking. What is the mechanism for fine-tuning, and by virtue of what would such fine-tuning constitute an improvement? I will confine my attention to human cognition, leaving aside the more theoretical question of how such fine-tuning might work in other cognitive agents. The question of how this works in human beings is a psychological question, and we can make only a limited amount of progress by speculating from our armchair. Despite this disclaimer, it may be useful to look at these issues at least briefly.

First consider what the fine-tuning might consist of. It will enable the mechanisms for generating state liking to take account of the agent's acquired knowledge of the world. The obvious suggestion is that this will consist of acquiring new conative dispositions responsive to beliefs involving acquired concepts. And the obvious mechanism for acquiring new conative dispositions is some form of conditioning. Thus one might become conditioned to like one's state better when one believes one is living in a democracy. It is clear that such complex beliefs about our current situation can affect our state liking. But, contrary to first appearances, it is not clear that the mechanism for this involves the acquisition of new conative dispositions through conditioning. First, let us consider some cases that, it seems, cannot be handled in terms of conditioning. We have "anticipatory emotions" like fear that some bad thing is going to befall us. This is fear of a future situation for which we have a negative feature liking. Such a fear can have profound effects on our present state liking. Similarly, our present state liking can be enhanced by pleasurable anticipation of something good that we expect to happen. Again, the sense in which it is good is that we have a positive feature liking for it. State liking is also effected by retrospective beliefs. I can bask in my past accomplishments or good fortune, and I can feel shame about my past transgressions or failures. These psychological (emotional) states are built-in mechanisms whereby current state liking is influenced by feature likings that we believe to have past or future instantiations.

In these examples, complex beliefs affect state liking indirectly by being beliefs about the instantiation of features for which we have feature likings. Feature likings are established by conditioning, but feature likings are logically different from conative dispositions. Feature likings encode *cumulative* state liking, whereas conative disposition produce *momentary* state liking. Insofar as feature likings affect state likings, they do so through the mediation of some other mechanism. Such mechanisms are provided by various sorts of prospective and retrospective emotions. However, these seem to work in terms of built-in conative dispositions. For instance, we have a built-in conative disposition to dislike being afraid. So these cases are handled, not by acquiring new conative dispositions, but by acquiring new feature likings that plug into built-in conative dispositions in various ways.

We have seen how feature likings believed to have prospective or retrospective instantiations can affect state liking. What about features likings believed to have present instantiations? These feature likings do not automatically affect our state liking. For example, I like living in a democracy, and I like having a good mountain bike. Similarly, I dislike it that certain politicians are currently in office. But most of the time, I do not think about these aspects of my life and they do not affect my state liking. On the other hand, sometimes I do focus my attention on the features, and then they do enhance or detract from my state liking. In the case of such currently instantiated feature likings, *reflecting on* the feature elicits a conative response. We can think of this as another built-in conative disposition.

Apparently feature likings play a dual role. They are the denizens of the evaluative database, to which practical deliberation appeals in deciding what actions to perform. But with the mediation of built-in conative dispositions concerning prospective and restrospective emotions and conative dispositions concerning reflecting on currently instantiated feature likings, they achieve the effect of acquired conative dispositions without our actually acquiring new conative dispositions. The feature liking is encoded as a conative response to thinking about the feature, and that conative response is simultaneously used in decision-theoretic reasoning and in altering state liking.

To recapitulate, we began by looking for mechanisms for creating new conative dispositions, and that led us to look at feature likings and investigate how they are involved in state liking. My conclusion was that their involvement is via built-in conative dispositions that respond to the thought that features for which one has a feature liking are, have been, or will be instantiated. It seems fairly clear that if we have explicit beliefs about the value of a feature, and we have the thought that the feature is, has been, or will be instantiated, this can affect state liking similarly. We can diagram the multiple inputs to conative dispositions as in figure 4.2.

### 3.4 Fine-Tuning State Liking

It is fairly clear that feature likings, and also beliefs about the values of features, have the power to affect state liking using mechanisms like the above. We might put it by saying that feature "valuations" are *infectious*. But now let us ask what this accomplishes. In what sense does this fine-tune state liking to make it better serve the design goals of the cognitive architecture?

Conditioning is inherently short-sighted. Long term consequences tend to be ignored by conditioning, which focuses on current state liking. The infectiousness of feature likings can be viewed as a mechanism for circumventing this. Thinking about the long-term consequences of what I am doing can affect my current state liking and thereby contribute to the conative conditioning of current actions. Perhaps even more important is the fact that we do not have to be thinking about what the long term consequences will definitely be. Worrying about what they might be, fearing what we expect them to be, and taking pleasure in what we think we may accomplish, all affect current state liking and contribute to the conative conditioning of my current actions. This is different from post hoc conditioning because it occurs while the action is being performed, but it achieves similar results.

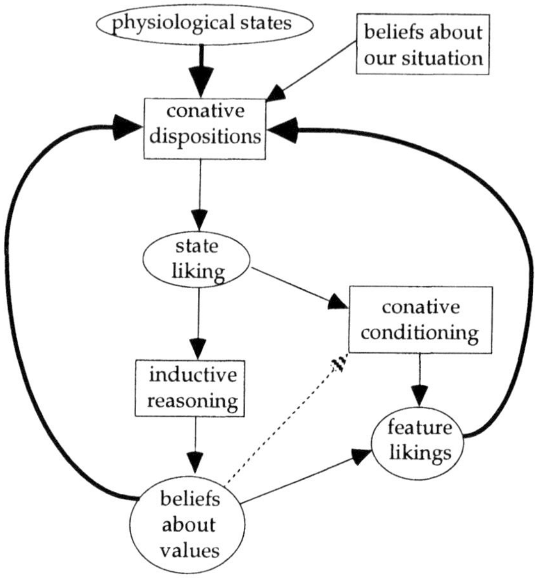


Figure 4.2. Multiple inputs to conative dispositions

We might call this *prospective conditioning*. It has the effect of bringing conditioned feature likings into line with the actual abstract values.

This explains why we are so constructed that thinking about (possible) future instantiations of features for which we have feature likings affects our current state liking. Similar observations can be made about why thinking about past instantiations should also affect our state liking. Think about reflecting on a stupidly impolite remark you once made. That has the power to make you uncomfortable now, and it certainly contributes to preventing you from repeating your *faux pax*. Thus it seems to play a role in the conative conditioning of your earlier action. We might call this *retrospective conditioning*.

These considerations indicate that conative conditioning has a fairly rich structure. We tend to think of conditioning as a low-level process that is phyologenetically earlier than higher cognition and independent of it. That is, it is cognitively impenetrable. This may be right for classical and operant conditioning, but it does not seem to be true of conative conditioning. It was observed above that conative conditioning does not have the same general profile as either classical or operant conditioning. Now it is emerging that it also seems to make heavy use of the results of high-level cognition. A general problem for conditioning is allocating credit or blame. You do a number of things, and then you feel good or bad. Conditioning is supposed to use this to attach a conative response to your actions. But ideally only the actions that are causally involved in making you feel good or bad should be conditioned. Conative conditioning seems to be sensitive to your beliefs about these causal processes. Clearly, post hoc conditioning is entirely dependent on such beliefs. It is based on your thoughts about what the consequences of an action *might have been*. Retrospective conditioning is also dependent on such beliefs, because it is dependent on your thoughts about what the consequences of a past action *are*. For example, I believe that it is my thinking about my *faux pax* that is making me currently uncomfortable. Similarly, prospective conditioning depends on my thinking about what the consequences of my current action might be or are likely to be.

It is this rich structure of conative conditioning and the fact that it is not cognitively impenetrable that makes the infectiousness of feature likings useful. Cruder forms of conditioning would not profit from it. This is why feature likings can be regarded as fine tuning innate conative dispositions.

### 3.5 *"If it feels good, keep doing it"*

It was observed in chapter 3 that feature likings for activities enter into a Q&I module for assessing the expected values of the activities, and that often provides an important shortcut for practical decision making. There is an interesting variant of this shortcut that appeals to state liking directly without going through conditioned feature likings. It involves a rule of thumb for decision making that has roughly the form, "If it feels good, keep doing it." This is about continuing actions rather than choosing actions. It can be a surprise that we are enjoying an activity. Choosing an autobiographical example, salesmen in bike shops are well known for having "attitude problems". They are often more interested in impressing the customer with their knowledge than in helping the customer. I went into a bike shop to talk to a particular salesman that I knew could supply me with information that I wanted, but I did not expect to enjoy talking to him. Midway through our conversation, I realized that I was enjoying it, and was prolonging the discussion more than I had planned. He was acting friendly and taking me seriously. At the time I wondered what the causal mechanism was that led me to prolong the conversation. I certainly had not engaged in any decision-theoretic reasoning about it. It now appears to me that it was this rule of thumb that led me to prolong the conversation. This rule of thumb is a partial substitute for full-fledged decision-theoretic reasoning. Note that to apply it, we must know what it is that feels good. Again, high-level cognition is involved. The fact that we are doing something and we like our current state does not indicate that this activity is responsible for the state liking. For example, I may be driving a car and talking to a friend at the same time. I may find the driving tedious, but enjoy talking to my friend. How do I sort that out? Perhaps by seeing how the time course of state liking fluctuates with fluctuations in the activities.

We also choose actions by *remembering* how they felt and noting that they felt good. We might think of this as inductive reasoning about abstract values based on a single case. I noted above that although I am aware that I enjoy skiing whenever I go, I have difficulty motivating myself to go. Perhaps my problem is that I do not remember how it felt. This may be because there is no general feeling associated with skiing—just lots of diverse feelings that are good on sum but hard to encapsulate in imagination or memory.

We also employ a converse of the above rule of thumb, viz., "If it feels bad, stop doing it." By appealing to the infectiousness of feature liking, this can help us take account of long term consequences. For example, I dislike going to the dentist. However, when I reflect upon the long term consequences of not going, I find that I dislike them even more. I imagine having tooth decay and gum rot, and that affects my current state liking, with the result that not going "feels bad", and hence I am inclined to refrain from not going. That is, this inclines me to go to the dentist.

## 4. Conclusions

This chapter has consisted of a number of miscellaneous observations about evaluative cognition. They are not an integral part of the theory of the book. But they interest me, so I have included them. By and large they concern highly contingent features of human cognition. These play a role in making human cognition efficient, and they also explain some of the more peculiar aspects of human evaluative cognition. In particular, they highlight the fact that humans find it difficult to employ explicit knowledge of expected values to override conditioned features likings, but they also explain why it would be difficult to build an agent that was not subject to this same difficulty. It also emerges that there are complex interconnections between state likings and feature likings, and they depend heavily on the cognitive penetrability of conative conditioning.

# 5
# The Database Calculation

Chapters 2–4 sketched a theory of evaluative cognition according to which values are encoded in feature likings, and the latter are stored in an evaluative database. The evaluative database provides efficient storage because it assumes defeasibly that values are additive, and hence the values of many compound features can be computed by summing the values of their constituents. This computation was called "the database calculation". This chapter completes the account by giving a precise characterization of the database calculation, investigating its logical credentials, and examining how it can be used for computing expected values.

I do not suppose that cognizers implement the database calculation by doing explicit mathematical calculations. Rather, the calculations go on in the background, performed by the cognitive system rather than the agent. I suggested in chapter 4 that it is very likely that they are implemented in humans by something like a connectionist network that also takes account of various parameters of the present circumstances in computing feature likings.

In section 1, the database calculation will be formulated precisely. This is to treat it as a principle for storing values in the evaluative database. In cases where values do not conform to the database calculation, they must be stored explicitly. For this to achieve compactness, it must be the case that values generally do conform to the database calculation. In section 2, it will be shown that general principles of causal reasoning make it defeasibly reasonable to expect values to conform to the database calculation, thus justifying its use. The connection between causal reasoning and the database calculation derives from the fact that the concept of value developed in chapter 3 is itself causal. Roughly, the value attached to a feature purports to be a measure of the tendency of the feature to cause state liking.

In humans, values are measured by analog representations rather than real numbers, but I presume that the analog representations and the computations on them constitute an approximation to arithmetic on the real numbers. So I will adopt the fiction of supposing that the evaluative database stores real-valued measures of abstract values. The database calculation as performed in human beings should approximate the real-valued computation described here.

## 1. The Database Calculation

It was argued in chapter 2 that the only way for an agent to have access to the large number of values required by evaluative cognition is by explicitly storing only a very small subset of the values, relying upon computation to retrieve those not explicitly stored. This computation was called "the database calculation". This section makes the database calculation precise.

The proposal is to store only those values that cannot be computed on the basis of other stored values. Values attach to features (types of situations). Some features can be regarded as built out of others, and I represent that with conjunction. If $P$ contains $Q$ as a constituent, then $P$ is a conjunction of features and $Q$ is one of its conjuncts. This imposes rather narrow constraints on the logical form of the features having values stored in the evaluative database. They cannot, for example, be disjunctions of one another. The justification for this restriction will be that expected values are computed by computing the values and probabilities of "scenarios", which are conjunctions of value-laden features. So only the latter need to be evaluated by the database calculation.

Value computations for conjunctive features will be based upon numerical values stored in an evaluative database of "computationally primitive values". The features having stored values will be said to be "primitively value-laden". Because the value of a feature can change when it is combined with other features (changes of context), some of these primitively value-laden features will be conjunctions having others as conjuncts. For example, there may be one value assigned to *eating a bowl of vanilla ice cream*, and another value assigned to *eating a bowl of vanilla ice cream shortly after eating a dill pickle.* The primitively value-laden features will also be allowed to contain free variables, corresponding to parameters that affect the value, in which case the value associated will be a function of those free variables rather than a constant value.

Features occurring in the evaluative database that are not equivalent to conjunctions of other features to which the evaluative database also assigns values will be called "value-theoretically simple" features. That a feature is value-theoretically simple is not a comment on the metaphysical structure of the world, but just a comment about the structure of our evaluative database. It tells us only that the feature is not a conjunction of simpler features assigned values by the evaluative database.

**U**($P$), the abstract value of a feature $P$, is the mathematical expectation of the amount of cumulative state liking caused by being in $P$. Two features $P$ and $Q$ are *value-theoretically independent* iff $\mathbf{U}(P\&Q) = \mathbf{U}(P)+\mathbf{U}(Q)$. The intent is that the evaluative database will contain a feature that is a conjunction only when a failure of value-theoretic independence prevents the value of the conjunction from being computable from the values of some of its conjuncts. To make this precise, we must state the rules for computing values for features that are not assigned values directly.

I will take conjunctions to be of arbitrary length, and the conjuncts of a conjunction $(P_1\&...\&P_n)$ will be $P_1,...,P_n$. A *subconjunction* of $(P_1\&...\&P_n)$ will be any single conjunct or conjunction of some of the conjuncts of $(P_1\&...\&P_n)$. Subconjunctions are conjunctions of subsets of conjuncts. Note that $(P_1\&...\&P_n)$ is one of its own subconjunctions. It is also convenient to identify the conjunction of two conjunctions with the conjunction of their conjuncts.

Turning to the rules for computing values for features that are not assigned values directly, the simplest case is a feature $S$ that has no primitively value-laden subconjunction. This signifies that the feature is value-neutral, that is, $\mathbf{U}(S) = 0$.

Consider a feature $(P_1\&P_2)$ where $P_1$ and $P_2$ are value-theoretically simple features. If $P_1$ is assigned a value $\mathbf{U}(P_1)$ directly, and $P_2$ is not assigned a value, then $\mathbf{U}(P_2) = 0$ and the presumption is that $P_1$ and $P_2$ are value-theoretically independent, in which case $\mathbf{U}(P_1\&P_2) = \mathbf{U}(P_1)+\mathbf{U}(P_2) = \mathbf{U}(P_1)$. This default computation will only be overridden if $(P_1\&P_2)$ is assigned a value directly.

Consider a feature $(P_1\&...\&P_n)$ where $P_1,...,P_n$ are value-theoretically simple features. If just one subconjunction of $(P_1\&...\&P_n)$ is assigned a value, then as above that will also be the value computed for $(P_1\&...\&P_n)$. But suppose instead that several subconjunctions are assigned values. We can distinguish several cases:

(a) Let us define:

$(A_1\&...\&A_n)$ *includes* $(B_1\&...\&B_m)$ iff $\{B_1,...,B_m\} \subseteq \{A_1,...,A_n\}$.

There might be one primitively value-laden subconjunction $S$ including all other primitively value-laden subconjunctions of $(P_1\&...\&P_n)$. Then the value assigned to $S$ will override the values assigned to any other subconjunctions, and so it will be computed that $\mathbf{U}(P_1\&...\&P_n) = \mathbf{U}(S)$.

(b) There might be several primitively value-laden subconjunctions $S_1,...,S_k$ such that (i) every primitively value-laden subconjunction of $(P_1\&...\&P_n)$ is included in some $S_i$, and (ii) the $S_i$'s have no conjuncts in common. This should signify that the $S_i$'s are value-theoretically independent and hence we should have $\mathbf{U}(P_1\&...\&P_n) = \mathbf{U}(S_1)+...+\mathbf{U}(S_k)$.

(c) Suppose $(P_1\&...\&P_n)$ has two primitively value-laden subconjunctions $S_1$ and $S_2$ neither of which includes the other, but suppose $S_1$ and $S_2$ have the conjunct $P$ in common. To identify $\mathbf{U}(P_1\&...\&P_n)$ with $\mathbf{U}(S_1)+\mathbf{U}(S_2)$ is to double count any contribution from $P$. If $\mathbf{U}(P) = 0$, this is not a problem, but what happens when $\mathbf{U}(P) \neq 0$? Notice that this is not a substantive question about value. Rather, it is a question about how best to organize the evaluative database in the interest of compactness. We could simply rule that in this case $(P_1\&...\&P_n)$ must be assigned a value directly, but additional compactness can be achieved by making some default assumptions. The value of a feature $P$ in a context $C$ may be different from the value of $P$ simpliciter. Let us write this as $\mathbf{U}(P/C)$. We can identify this with the value $P$ contributes to the conjunction $(P\&C)$ over and above the value of $C$, and define it precisely as follows:

$$\mathbf{U}(P/C) = \mathbf{U}(P\&C) - \mathbf{U}(C).$$

Then we can extend the presumption of value-theoretic independence by assuming that $\mathbf{U}(P_1\&...\&P_n) = \mathbf{U}(S_1/P) + \mathbf{U}(S_2/P) + \mathbf{U}(P) = \mathbf{U}(S_1) + \mathbf{U}(S_2) - \mathbf{U}(P)$.

To illustrate, suppose we want to compute the value of eating a banana split, where that consists of vanilla ice cream, bananas, and chocolate sauce. Let us suppose that the value of eating vanilla ice cream with bananas is greater than the sum of the values of eating vanilla ice cream and bananas separately, and similarly for the value of eating vanilla ice cream and chocolate sauce. Then the value of eating the banana split is the value of eating vanilla ice cream plus the value added to that by the bananas plus the value added to that by the chocolate sauce. That in turn is the value of eating vanilla ice cream with bananas plus the value of eating vanilla ice cream with chocolate sauce minus the value of eating vanilla ice cream by itself.

There is a useful way to reformulate the preceding. Identifying conjunctions with the sets of their conjuncts, $S_1 \cap S_2$ is the conjunction of the conjuncts $S_1$ and $S_2$ have in common. Then $\mathbf{U}(P) = \mathbf{U}(S_1 \cap S_2)$, so the preceding computation amounts to positing that

$$\mathbf{U}(P_1\&...\&P_n) = \mathbf{U}(S_1) + \mathbf{U}(S_2) - \mathbf{U}(S_1 \cap S_2).$$

(d) More generally, suppose $(P_1\&...\&P_n)$ has several primitively value-laden subconjunctions $S_1,...,S_k$ none of which is included in any other, but some of which have conjuncts in common. How do we generalize the preceding calculation? To get the mathematics right, consider an analogous problem that is better understood. Computing the value of a conjunction of features constructed out of several overlapping value-laden subconjunctions should be analogous to computing the volume of a three-dimensional figure constructed out of several overlapping figures. Suppose $X$ is a three-dimensional figure composed of the sometimes overlapping figures $X_1,...,X_n$. This is illustrated by figure 5.1, where the composite figure consists of a rod passing through a block. We can identify a three-dimensional figure with the set of its points in a three-dimensional space. Then $X = X_1 \cup ... \cup X_n$. Volume is an additive set function, in the sense that

$$vol(X \cup Y) = vol(X) + vol(Y) - vol(X \cap Y).$$

It follows from this that

$$vol(X) = vol(X_1) + vol(X_2 \cup ... \cup X_n) - vol((X_1 \cap X_2) \cup ... \cup (X_1 \cap X_n)).$$

The analogous theorem holds in general for additive set functions.

The computation we produced in (c) is similar to saying that the database calculation treats $\mathbf{U}$ as an additive set function. Identifying a conjunction

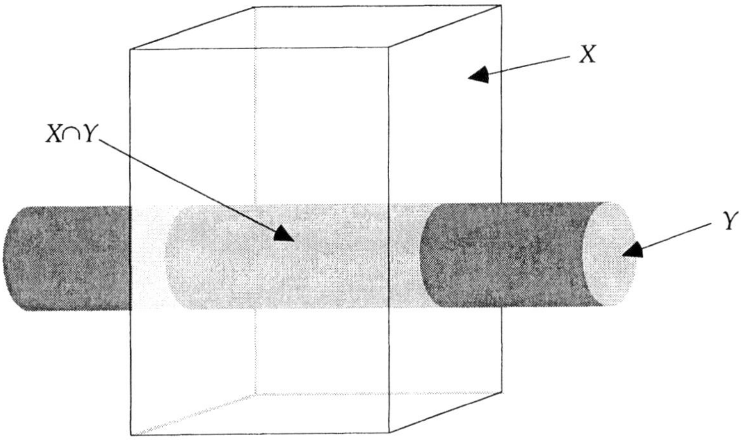


Figure 5.1. Composite three-dimensional figure

with the set of its conjuncts, the computation in (c) amounts to first identifying $\mathbf{U}(P_1\&\ldots\&P_n)$ with $\mathbf{U}(S_1 \cup S_2)$ and then computing that $\mathbf{U}(S_1 \cup S_2) = \mathbf{U}(S_1) + \mathbf{U}(S_2) - \mathbf{U}(S_1 \cap S_2)$. On analogy to the preceding principle about volumes, let us define the notion of the *additive utility* of a *set* of primitively value-laden subconjunctions. Symbolizing this as $\mathbf{AU}\{S_1,\ldots,S_k\}$, we can define recursively:

$$\mathbf{AU}\{S_1,\ldots,S_k\} = \mathbf{U}(S_1) + \mathbf{AU}\{S_2,\ldots,S_k\} - \mathbf{AU}\{S_1 \cap S_2,\ldots,S_1 \cap S_k\}.$$

The general form of the *database calculation* is then:

> Where $S_1,\ldots,S_k$ are the primitively value-laden subconjunctions of $(P_1\&\ldots\&P_n)$ none of which is included in any other, $\mathbf{U}(P_1\&\ldots\&P_n) = \mathbf{AU}\{S_1,\ldots,S_k\}$.

Equivalently, the database calculation treats $U$ as an additive set function down to the level of primitively value-laden conjunctions. For the latter the values are posited rather than computed.

Thus far I have represented the database calculation as a principle for retrieving values from the evaluative database. It is not a substantive principle about values, but rather an organizational principle for the evaluative database. If the actual value of $(P_1\&\ldots\&P_n)$ does not accord with this computation, then it must be included as a primitive entry in the database. On the other hand, there is a substantive assumption underlying this organization of the evaluative database. Organizing it in this way will only achieve compactness if the values of features are normally related in accordance with the database calculation. The justification of this assumption will be addressed next.

## 2. Justifying the Database Calculation

In the preceding discussion it was supposed that the evaluative database represented the complete and correct assignment of values to features. But that is an unrealistic assumption. The evaluative database is supposed to represent the cognitive agent's *stored information* about expected values of features, and that information can be wrong or incomplete. So let us assume instead that the evaluative database represents the expected values of features *to the best of the agent's knowledge*. When the agent knows the expected value of a conjunction and it cannot be computed from the expected values of its conjuncts in accordance with the database calculation, then the value is entered into the database as an explicit entry. If the value can be computed *or if the value is unknown*, then no entry is made in the database.

Organizing the database in this way has the consequence that the absence of an entry for a particular feature can mean either of two things. The value may be redundant, in the sense that it can be computed from other entries using the database calculation, or the value may be unknown. It may seem that this disjunctive significance makes the database useless for retrieving values for items for which there are no explicit entries. If there is no entry for a feature because we have no independent knowledge of its value, but one can be computed using the database calculation, why should we expect that to be the right value? As I will now argue, the power of the evaluative database arises from the fact that the database calculation can be assumed defeasibly to return the right value even when we have no explicit knowledge of the value to be retrieved. This is what makes the organizational principle efficient.

Recalling that $S_1 \cap S_2$ is the conjunction of the conjuncts $S_1$ and $S_2$ have in common, the database calculation can be grounded on two principles:

> ***Irrelevance***
> If $S$ is a feature and the agent does not know the utility of $S$ or of any subconjunction of $S$, it is defeasibly reasonable to take $\mathbf{U}(S)$ to be 0.
>
> ***Additivity***
> If $S_1,\ldots,S_k$ are subconjunctions of a feature $S$, the agent knows the utilities of each $S_i$ but does not know the utility of $S$ or of any subconjunctions including any $S_i$, then it is defeasibly reasonable to take $\mathbf{U}(S)$ to equal $\mathbf{AU}\{S_1,\ldots,S_k\}$.

I will argue that these two principles are grounded on principles that we commonly use for reasoning about causes.

Recall that our concept of value is an essentially causal concept. The value of a feature $S$, measured by $\mathbf{U}(S)$, is the expected value of the cumulative state liking caused by being in feature $S$. Irrelevance is a consequence of the following general principle:

> ***Causal Irrelevance***
> If the agent has no reason to think otherwise, it is reasonable to think that *P* does not cause *Q*.

That is, it is defeasibly reasonable to expect arbitrary features to be causally independent. This is a general principle of causal reasoning that we employ regularly. We base our causal reasoning on those considerations that we know to be relevant, and we assume defeasibly that other considerations will not disrupt the causal connections that we know about. Without this assumption, causal reasoning would be impossible. This has the consequence that if the agent has no reason to think otherwise, it is defeasibly reasonable to think that the obtaining of a feature *S* does not cause a change in state liking, and hence $\mathbf{U}(S) = 0$.

We can generalize the principle of causal irrelevance. Our causal knowledge is to the effect that certain causal processes operate in specific situations. A causal process can be viewed as a tree of causes and effects strung together into causal chains and originating from a single cause—the *origin* of the process. Each branch through the tree is a causal chain. This is diagrammed in figure 5.2. To say that a causal process *p operates in* a situation *S* is just to say that *S* occurs, the origin of the tree is contained in *S*, and *S* causes the effects of *p* in accordance with the causal chains constituting *p*.

Our knowledge of the current situation is always incomplete. On the basis of general causal knowledge we may expect that a causal process will operate in a particular situation that now obtains. But we can never conclusively rule out the possibility of there being something that will interfere with causal processes that we expect to occur on the basis of our limited knowledge. Thus if we are to be able to draw any conclusions about the causal consequences of an event, we must be able to assume defeasibly that

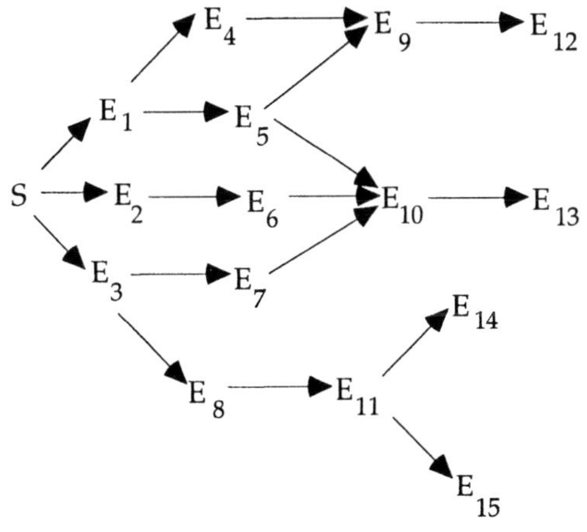


Figure 5.2. A causal process operating in *S*

if a causal process can be expected to operate in one situation, it can also be expected to operate in a more fully specified situation. If we could not make this assumption, causal reasoning would be impossible. Let us symbolize "$S_2$ is a more fully specified situation than $S_1$" (that is, $S_1$ is contained in $S_2$) as "$S_1 \lessdot S_2$". So the principle we employ is:

> "$S_1 \lessdot S_2$ and the causal process $p$ normally operates in (or does not operate in) $S_1$" is a defeasible reason for "The causal process $p$ normally operates in (or does not operate in) $S_2$".

Positive knowledge of causal processes takes precedence over negative knowledge:

> If $S_1$ and $S_2$ are situations neither of which includes the other, "The causal process $p$ normally operates in $S_1$ and the causal process $p$ does not normally operate in $S_2$" is a defeasible reason for "The causal process $p$ normally operates in $S_1$&$S_2$".

Furthermore, knowledge of how causal processes operate in a more inclusive situation takes precedence over knowledge how they operate in a less inclusive situation:

> If $S_1 \lessdot S_2 \lessdot S_3$ and different causal processes operate in $S_1$ than in $S_2$, then a defeasible inference to what causal processes normally operate in $S_3$ based upon knowledge of the causal processes normally operative in $S_2$ takes precedence over a defeasible inference based upon knowledge of the causal processes normally operative in $S_1$.

I will take the *Principle of Causal Independence* to be the conjunction of these three principles of defeasible causal reasoning.

I will now show how, given our causal concept of value, the principle of additivity can be derived from the principle of causal independence. The details are mathematically complex, so the reader who is willing to accept the result without wading through the mathematics is invited to skip the rest of this section.

If $S$ is a value-laden feature, then there is some probability that $S$ will cause various amounts of cumulative state liking. In a situation in which $S$ causes some amount of cumulative state liking, there is a set of causal processes operative in $S$ that causes cumulative state liking. The identity of a causal process is determined, in part, by what it causes. So whenever one of these causal processes is operative, it causes the same quantity of cumulative state liking. Accordingly, the cumulative state liking caused by $S$ is the sum of the cumulative state likings caused by the causal processes.

As defined in chapter 3, *cumulative-state-liking*($S$) is the cumulative state liking caused by $S$. Let **CP**($S$) be the set of value-producing causal processes operative in $S$. For each causal process $p$ in **CP**($S$), let *val*($p$) be the quantity of cumulative state liking produced by $p$. So what has just been argued is that *cumulative-state-liking*($S$) is the sum of the *val*($p$) for all $p$ in **CP**($S$). Let **AV**($X$) be the sum of the *val*($p$) for $p$ in a set $X$ of causal processes. Thus *cumulative-state-liking*($S$) =

**AV**(**CP**($S$)).

It follows immediately from the definition of **AV**($X$) that it is an additive set function, in the sense that:

$$\mathbf{AV}(X \cup Y) = \mathbf{AV}(X) + \mathbf{AV}(Y) - \mathbf{AV}(X \cap Y).$$

By the principle of causal independence, if $S_1 \prec S$ then it is defeasibly reasonable to expect all value-producing causal processes operative in $S_1$ to still be operative in $S$. In other words, $\mathbf{CP}(S_1) \subseteq \mathbf{CP}(S)$. More generally, if $S_1,...,S_k$ are value-laden subconjunctions of $S$, it is defeasibly reasonable to expect that $\mathbf{CP}(S_1) \cup ... \cup \mathbf{CP}(S_k) \subseteq \mathbf{CP}(S)$. By the third clause of the principle of causal independence, if $S^*$ is a value-laden subconjunction of $S$ that is included in one of the $S_i$'s, it is defeasibly reasonable to expect that a causal process in $\mathbf{CP}(S^*)$ is operative in $S$ only if it is operative in $S_i$, so subsumed subconjunctions make no contribution to $\mathbf{CP}(S)$. Furthermore, if a causal process is not operative in any of $S_1,...,S_k$, it is defeasibly reasonable to expect that it is not operative in $S$. Thus we can conclude defeasibly that:

> if $S_1,...,S_k$ are the value-laden subconjunctions of $S$ that are not included in any other value-laden subconjunctions then $\mathbf{CP}(S) = \mathbf{CP}(S_1) \cup ... \cup \mathbf{CP}(S_k)$.

Hence it is defeasibly reasonable to expect that *cumulative-state-liking*($S$) is $\mathbf{AV}(\mathbf{CP}(S_1) \cup ... \cup \mathbf{CP}(S_k))$.

Because **AV**($X$) is an additive set function, we can compute

$$\mathbf{AV}(\mathbf{CP}(S_1) \cup ... \cup \mathbf{CP}(S_k))$$

$$= \mathbf{AV}(\mathbf{CP}(S_1)) + \mathbf{AV}(\mathbf{CP}(S_2) \cup ... \cup \mathbf{CP}(S_k)) - \mathbf{AV}(\mathbf{CP}(S_1) \cap (\mathbf{CP}(S_2) \cup ... \cup \mathbf{CP}(S_k)))$$

$$= \mathbf{AV}(\mathbf{CP}(S_1)) + \mathbf{AV}(\mathbf{CP}(S_2) \cup ... \cup \mathbf{CP}(S_k)) \\ - \mathbf{AV}(\mathbf{CP}(S_1) \cap \mathbf{CP}(S_2) \cup ... \cup \mathbf{CP}(S_1) \cap \mathbf{CP}(S_k)).$$

Recalling that $S_i \cap S_j$ is the conjunction of the conjuncts $S_i$ and $S_j$ have in common, $S_i \cap S_j \prec S_i$ and $S_i \cap S_j \prec S_j$. Thus it is defeasibly reasonable to expect that $\mathbf{CP}(S_i \cap S_j) \subseteq \mathbf{CP}(S_i)$ and $\mathbf{CP}(S_i \cap S_j) \subseteq \mathbf{CP}(S_j)$, and hence $\mathbf{CP}(S_i \cap S_j) \subseteq \mathbf{CP}(S_i) \cap \mathbf{CP}(S_j)$. By the first two clauses of causal independence, it is defeasibly reasonable to expect that causal processes not operative in either $S_i$ or $S_j$ are not operative in $S_i \cap S_j$, so we can conclude defeasibly that

$$\mathbf{CP}(S_i \cap S_j) = \mathbf{CP}(S_i) \cap \mathbf{CP}(S_j).$$

Combining this with the above computation, we can conclude defeasibly that

$$\mathbf{AV}(\mathbf{CP}(S_1) \cup ... \cup \mathbf{CP}(S_k))$$

$$= \mathbf{AV}(\mathbf{CP}(S_1)) + \mathbf{AV}(\mathbf{CP}(S_2) \cup ... \cup \mathbf{CP}(S_k)) - \mathbf{AV}(\mathbf{CP}(S_1 \cap S_2) \cup ... \cup \mathbf{CP}(S_1 \cap S_k)).$$

Let us define the *additive-value* of a set of features as follows:

$$\textit{additive-value}\{S_1,\ldots,S_k\} = \mathbf{AV}(\mathbf{CP}(S_1)\cup\ldots\cup\mathbf{CP}(S_k)).$$

Then the result of the preceding computation can be rewritten as:

$$\begin{aligned}&\textit{additive-value}\{S_1,\ldots,S_k\}\\ &= \textit{cumulative-state-liking}(S_1) + \textit{additive-value}\{S_2,\ldots,S_k\}\\ &\qquad - \textit{additive-value}\{S_1\cap S_2,\ldots,S_1\cap S_k\}.\end{aligned}$$

It was argued above that it is defeasibly reasonable to expect that *cumulative-state-liking*($S$) = *additive-value*$\{S_1,\ldots,S_k\}$, so we can conclude defeasibly that

$$\begin{aligned}&\textit{cumulative-state-liking}(S)\\ &= \textit{cumulative-state-liking}(S_1) + \textit{additive-value}\{S_2,\ldots,S_k\}\\ &\qquad - \textit{additive-value}\{S_1\cap S_2,\ldots,S_1\cap S_k\}.\end{aligned}$$

This looks much like the database calculation, except that the database calculation is about the mathematical expectations of values-caused rather than about values-caused themselves. We can, however, derive the database calculation from this result. The above computation of *additive-value* can be expanded recursively to yield the result that

$$\begin{aligned}&\textit{cumulative-state-liking}(S) =\\ &\textstyle\sum_{1\leq i\leq n}\textit{cumulative-state-liking}(S_i) - \sum_{i\neq j}\textit{cumulative-state-liking}(S_i\cap S_j)\\ &\qquad + \textstyle\sum_{i\neq j, i\neq k, j\neq k}\textit{cumulative-state-liking}(S_i\cap S_j\cap S_k) - \ldots\end{aligned}$$

$\mathbf{U}(S)$ is the mathematical expectation of *cumulative-state-liking*($S$). The mathematical expectation of a sum of functions is the sum of their mathematical expectations, so it follows that we can defeasibly expect:

$$\mathbf{U}(S) = \mathbf{U}(S_1)+\ldots+\mathbf{U}(S_k) - \textstyle\sum_{i\neq j}\mathbf{U}(S_i\cap S_j) + \sum_{i\neq j, i\neq k, j\neq k}\mathbf{U}(S_i\cap S_j\cap S_k) - \ldots$$

We can similarly expand the definition of additive utility to obtain:

$$\begin{aligned}&\mathbf{AU}\{S_1,\ldots,S_k\} =\\ &\mathbf{U}(S_1)+\ldots+\mathbf{U}(S_k) - \textstyle\sum_{i\neq j}\mathbf{U}(S_i\cap S_j) + \sum_{i\neq j, i\neq k, j\neq k}\mathbf{U}(S_i\cap S_j\cap S_k) - \ldots\end{aligned}$$

Thus we can defeasibly expect that $\mathbf{U}(S) = \mathbf{AU}\{S_1,\ldots,S_k\}$. This is the principle of Additivity.

My conclusion is that defeasible principles of causal reasoning make it defeasibly reasonable to expect the database calculation to hold in any specific case. This justifies the use of the evaluative database in decision-theoretic reasoning.

## 3. Feature-Based Evaluative Cognition

Cognitive agents implement the doxastic-conative loop. For now I am assuming that the implementation of the doxastic-conative loop requires the decision-theoretic evaluation of actions. I will refine that assumption in subsequent chapters. The decision-theoretic evaluation of actions must proceed in terms of an evaluative database assigning a cardinal measure of values to concepts that pick out value-laden features. I have argued that computational feasibility requires that most of the abstract values employed in evaluative cognition must be derived, using the database calculation, from simpler values stored explicitly in the database. I will refer to this overall picture as *feature-based evaluative cognition.* The rest of the book is devoted to giving an account of how these stored and computed values are to be used in decision making. Part II investigates the probabilities that are required for computing expected values, and then Part III investigates how expected values should be used in rational decision making.

# Part II

# Probabilities

# 6
# Subjective Probabilities

## 1. Two Kinds of Probabilities

The optimality prescription is formulated in terms of expected values. Expected values are defined in terms of probabilities and utilities, and require a real-valued measure of each. Part I of the book investigated the source of the real-valued measure of utilities. Now we must ask about the probabilities. There are two general approaches to understanding probabilities. The preponderance of contemporary work on the optimality prescription assumes what is known as *subjective expected utility theory*. According to subjective expected utility theory, the probabilities employed in rational decision making should be subjective probabilities. An agent's *subjective probability* for a proposition *P* is supposed to be the degree of belief that the agent ought to have in *P* (relative to his current epistemic situation). Here it is assumed that belief comes in degrees. More contentiously, it is assumed that degrees of belief satisfy, or ought to satisfy, the probability calculus. This claim is generally defended by appeal to the "Dutch book argument", which will be discussed below.

Subjective probabilities are supposed to be measures of an agent's degrees of belief. As such, they represent facts about the agent—not about the world. This can seem problematic in the context of rational decision making. In deciding what to do, it seems that we want to take account of how likely various outcomes are, where these likelihoods reflect actual probabilities in the world. Thus instead of defining expected values in terms of subjective probabilities, we might want to define them in terms of probabilities that are somehow "more objective". Theories of *objective probability* attempt to provide foundations for such reasoning. They are the topic of chapter 7.

This chapter will be devoted to an examination of subjective probabilities. Historically, theories of objective probability came first. Subjective probability theory arose as a response to difficulties encountered in making sense of objective probabilities. However, I will argue that there are even more serious difficulties for making sense of subjective probabilities in a way that makes them useful in a theory of practical cognition. Thus I will urge a return to objective probabilities.

## 2. Subjective Probabilities and Degrees of Belief

Theories of subjective probability begin by observing that cognitive agents hold beliefs with varying degrees of conviction. Furthermore, as one's degree of conviction increases, it seems that one should become more willing to bet that one is right. To get a numerical measure of degrees of belief, the standard subjectivist strategy is to cash it out in terms of gambles. We talk about making a bet at certain "odds". If you *accept* a bet offered to you at $r{:}n$ odds and you wager $x$ dollars, that means that if you win the bet you receive a payoff of $rx$, but if you lose you must pay out $nx$. If you *offer* a bet at $r{:}n$ odds, the amounts are reversed. When we talk about accepting a bet at $r{:}n$, we are talking about a bet that is offered to you. In comparing odds, we say that they are *better* (from the perspective of being offered to you) if the ratio of $r$ to $n$ is higher. So better odds correspond to gambles with higher expected values.

Higher degrees of belief make it reasonable to accept bets with poorer odds. If I think that today is Tuesday, but I am not sure, I might be willing to accept a bet with 1:2 odds that I am right. That is, I would accept a bet on which I receive some value $x$ if today is Tuesday, but must pay $2x$ if I am wrong. On the other hand, if I am quite certain that today is Tuesday, I might be willing to accept a bet with 1:9 odds.

If the probability that $P$ is true is $p$, the expected value of betting $x$ dollars that $P$ is true at odds $r{:}n$ is $prx - (1 - p)nx$. Assuming the optimality principle, it is rational to accept such a bet (in preference to not betting) iff the expected value of accepting the bet is greater than 0. This holds iff $p > n/(n+r)$. This suggests the following definition of "degree of belief":

> A cognizer $S$ has *degree of belief* $n/(n+r)$ in a proposition $P$ iff $S$ would accept any bet that $P$ is true with odds better than $r{:}n$, but would not accept a bet with poorer odds.

Thus what the subjectivist *means* by "degree of belief" is something measured by betting behavior.

Subjective probabilities are sometimes identified with degrees of belief. On this construal, the subjective probability of $P$ for $S$ is simply $S$'s degree of belief in $P$. But there is a problem for this approach. It is generally agreed that the actual degrees of belief of real agents do not conform to the probability calculus. We can think of the probability calculus as being defined by Kolmogorov's axioms (Kolmogoroff 1933), which tell us that probability is an additive measure normalized to the interval between 0 and 1. In mathematical probability theory, probabilities are taken to attach to "states of affairs", but for their use in philosophy, we want to talk instead of the probability of a proposition being true. For our purposes it seems best to identify states of affairs with sets of logically equivalent propositions. Then we can axiomatize the probability calculus as follows:

THE PROBABILITY CALCULUS:

(1) PROB($P$ & ~$P$) = 0.

(2) PROB($P \vee$ ~$P$) = 1.

(3) PROB($P \vee Q$) = PROB(P) + PROB($Q$) – PROB($P$ & $Q$).

(4) If $P$ and $Q$ are logically equivalent then PROB($P$) = PROB($Q$).

In standard mathematical treatments of probability theory, it is common to add a fifth axiom. This is the principle of *countable additivity*, which says that if $P$ is equivalent to the disjunction of infinitely many disjuncts $Q_i$, where for any $i \neq j$, PROB($Q_i$ & $Q_j$) = 0, then PROB($P$) is the sum of the PROB($Q_i$). This axiom is adopted more to make the mathematics run smoothly than because it has a clear philosophical justification.[1] Countable additivity has been either rejected or questioned by most of the important writers in the foundations of probability theory, including de Finetti (1937), Reichenbach (1949), Jeffrey (1983), Skyrms (1980), Savage (1954), and Kyburg (1974). To the best of my knowledge, no one has given a clear counterexample to countable additivity, but here is one that seems to work. Let us suppose that the universe is of infinite spatial extent. We can imagine a theory of quantum gravity predicting that as energy is sucked into a black hole, a singularity is created in space from which photons are emitted spontaneously, but the location of the singularity is completely unpredictable. That is, the probability of the singularity being created in any region of space is the same as the probability of its being created in any other region of equal volume. More precisely, if $V_1$ and $V_2$ are regions of equal volume, then PROB(singularity created in $V_1$/singularity created in either $V_1$ or $V_2$) = PROB(singularity created in $V_2$/singularity created in either $V_1$ or $V_2$). Now divide space into disjoint cubes one meter on a side. As there are infinitely many of them, it follows that the probability of the singularity being created in any one cube is 0, but the probability of its being created in some cube or other is 1. Thus countable additivity fails. On the strength of examples like this, I will follow most philosophers and reject countable additivity.[2]

It is customary to introduce the following as a technical definition: A cognizer's degrees of belief are said to be *coherent* if they conform to the probability calculus, and *incoherent* otherwise. For computing expected values, it is normally assumed that the probabilities the decision maker assigns to outcomes conform to the probability calculus. But why should they? In particular, if probabilities are understood as degrees of belief, why should they be coherent in this sense? There is a standard argument, due originally to Frank Ramsey (1926), which is used to defend the claim that a cognizer

---

[1] Unfortunately, it is also assumed in some philosophical treatments of probability, e.g., Kaplan (1996) and Joyce (1998).

[2] I suspect that what this example really ilustrates is that probabilities should be measured using hyperreals (the real numbers supplemented with infinitesimals). Then we can say that the probability of the singularity being created in a particular cube is infinitesimal rather than zero. Something like countable additivity might then hold. It just does not hold for real-valued probabilities.

*ought* to have coherent degrees of belief. This is the Dutch book argument.[3] In betting parlance, a "Dutch book" is a combination of bets on which a person will suffer a collective loss no matter what happens. For instance, suppose you are betting on a coin toss and are willing to accept odds of 1:2 that the coin will land heads and are also willing to accept odds of 1:2 that the coin will land tails. I could then place two bets with you, betting 50 cents against the coin landing heads and also betting 50 cents against the coin landing tails, with the result that no matter what happens I will have to pay you 50 cents on one bet but you will have to pay me $1 on the other. In other words, you have a guaranteed loss—Dutch book can be made against you.

In order to maximize the plausibility that real agents will have degrees of belief, I have defined degrees of belief in a slightly different way than is customary. Let us say that a cognizer's degrees of belief are *symmetric* iff, for any proposition $P$, if his degree of belief in $P$ is $p$, then his degree of belief in $\sim P$ is at least $(1 - p)$. The standard subjectivist definition of "degree of belief" simply builds in the requirement that degrees of belief be symmetric by defining:

> A cognizer $S$ has degree of belief $n/(n+r)$ in a proposition $P$ iff $S$ would accept any bet that $P$ is true with odds better than $r{:}n$, and $S$ would accept any bet that $P$ is false with odds better than $n{:}r$.

One could reasonably question, however, whether degrees of belief, so defined, exist for all cognizers. Mightn't a cognizer simply refuse to bet unless he thinks he has a considerable advantage? This is one kind of risk aversion. So I have chosen the weaker definition. But this complicates the Dutch book argument. It must now have two stages. We must first argue that it is irrational to have non-symmetric degrees of belief, and then argue that if an agent's degrees of belief are symmetric then it is irrational to have degrees of belief that do not conform to the probability calculus.

If a cognizer's degrees of belief are not symmetric, then he could be offered the opposite of a Dutch book—a combination of bets on which he is guaranteed to have a collective win no matter what happens—that he would not accept. To see this, suppose $S$'s degree of belief in $P$ is $r$ and his degree of belief in $\sim P$ is $s$, where $s < 1 - r$. Let $\varepsilon = (1 - r) - s$. If $S$ is offered the opportunity to bet $1 that $P$ is true at odds $(r + \varepsilon/2){:}(1 - r)$ and bet $1 that $P$ is false at odds $(s + \varepsilon/2){:}(1 - s)$, he will decline both bets. But had he taken the bets then regardless of whether $P$ is true, he would have received a cumulative payoff of $\varepsilon/2$. The subjectivist can urge that it would be irrational to fail to accept a combination of bets on which one is guaranteed to profit, so it cannot be rational to have non-symmetric degrees of belief.

If a cognizer has symmetric degrees of belief, then the standard Dutch book argument can be brought to bear. This consists of a mathematical proof that if an agent has symmetric degrees of belief but they do not

---

[3] See also de Finetti (1937), Kemeny (1955), Lehman (1955), and Shimony (1955).

conform to the probability calculus then Dutch book can be made against him. That is, he will accept a combination of bets on which he is guaranteed to suffer a collective loss no matter what happens. It is alleged that it is irrational to put oneself in such a position, so it cannot be rational to have incoherent degrees of belief.

On the other hand, it is agreed by all that real cognizers do not have coherent degrees of belief. So instead of identifying subjective probabilities with a cognizer's actual degrees of belief, subjective probability theorists typically identify them with the degrees of belief the cognizers rationally *should* have. The Dutch book argument is then taken to establish that subjective probabilities, as rational degrees of belief, satisfy the probability calculus.

What should we make of the Dutch book argument? Suppose we agree that a cognizer ought to accept Dutch books and ought to avoid having Dutch books made against him. This has no implications about his betting behavior elsewhere unless we assume that he has degrees of belief in the technical sense of the definition. The definition makes a very strong requirement. It requires that if an agent has a degree of belief of $n/(n+r)$ in $P$, then he will accept *any* bet on $P$ at odds better than $r{:}n$, no matter what amount of money turns on the bet. Real agents are usually more inclined to accept small bets than large bets. This is a different sense in which they are risk averse. There may be *no* odds they are willing to accept regardless of the size of the bet, and that implies that they do not have degrees of belief. No argument has been given for the irrationality of risk aversion in this sense, but without that we have no reason to think that a cognizer rationally ought to have degrees of belief in the technical sense of the definition.

Even if it can be argued somehow that cognizers ought to have degrees of belief, we can still question what the lesson of the Dutch book argument should be. It is certainly undesirable to have Dutch book made against you, but is it truly irrational to have degrees of belief making that possible? Perhaps it is if we understand rationality in the sense of "ideal rationality", as discussed in chapter 1. However, real cognizers, with limited cognitive powers, will not be able to avoid having incoherent degrees of belief. For instance, if $P$ and $Q$ are logically equivalent propositions, but I am unaware of this and have different degrees of belief in them, Dutch book can be made against me. But it is not as if I am making a mistake in reasoning. I am just subject to unavoidable ignorance. No real agent can have knowledge of every logical equivalence. That is computationally impossible. Real rationality might prescribe *trying to find out* whether $P$ and $Q$ are logically equivalent, but that can take arbitrarily much reasoning, and it may be impossible to complete the reasoning before it is time to bet. So one cannot be rationally culpable for failing to discover the equivalence.

Because we cannot expect real agents to have coherent degrees of belief, no matter how diligent they are, identifying subjective probabilities with rational degrees of belief does not guarantee that they will conform to the probability calculus. But we only get this result because we are talking about real rationality rather than ideal rationality. We can avoid this difficulty by identifying subjective probabilities with "ideally rational degrees of belief" rather than actual degrees of belief. More accurately:

> *S*'s subjective probability for *P* is the degree of belief *S* (ideally) rationally should have in *P* (given his current epistemic situation).

Put another way, *S*'s subjective probability for *P* is the degree of belief *S* should have in *P* if, starting from his current epistemic state, he could perform all relevant reasoning. This is, in effect, the way subjective probabilities are normally defined, because subjective probability theorists usually mean "ideally rational" when they talk about rationality. But it must be emphasized that for non-ideal agents, subjective probabilities and rational degrees of belief can be different.

## 3. Belief Simpliciter

The theory of subjective probability is based on the notion of degrees of belief. This is to be contrasted with our ordinary notion of belief according to which we either believe or fail to believe something simpliciter. The orthodox subjectivist view is that our ordinary notion is confused, and such a binary notion of belief makes no sense. For instance, Richard Jeffrey (1970, 171–172) writes, "I am not disturbed that our ordinary notion of *belief* is only vestigially present in the notion of degree of belief. I am inclined to think that Ramsey sucked the marrow out of the ordinary notion, and used it to nourish a more adequate view." The rejection of binary beliefs is usually based on the assumption that if you believe something simpliciter, as opposed to having some qualified degree of belief in it, then you will simply assume it in your reasoning, without having to qualify it with a probability. But if you are willing to simply assume *P* for the purposes of deciding what to do, that seems to imply that if you are offered a wager on whether *P* is true, you should be willing to accept any odds. In other words, your degree of belief in *P* must be 1. However, it seems clear that, for at least for many of our beliefs, we would not be willing to bet absolutely anything on the belief being true. For example, I believe my car keys are on my dresser. But if you offer me a wager whereby I receive one penny if my car keys are there and must pay you one million dollars if they are not, I would probably decline the wager. The subjectivist concludes that I do not really believe, simpliciter, that my car keys are on my dresser. I just have a high degree of belief in that proposition.

To accommodate examples like the preceding, one might suggest that believing *P* simpliciter amounts merely to having a degree of belief that comes up to a certain threshold, where that threshold is less than 1. But this proposal runs afoul of the lottery paradox.[4] Pick any threshold you like, say .999999. Now suppose you hold a ticket in a fair lottery consisting of one million tickets. If you know that the lottery is fair, it seems you must believe

---

[4] The lottery paradox is due to Kyburg (1961). I follow Kaplan (1996) in my use of it here.

that some ticket will be drawn, but that each ticket individually has a probability of .999999 of not being drawn. On the threshold view of belief simpliciter, it would follow that you believe of each ticket that it will not be drawn. In fact, you may believe that initially. But that makes your set of beliefs inconsistent, because you also believe that some ticket will be drawn. As Kaplan (1996) observes, once you recognize this, you will probably retract your beliefs that the individual tickets will not be drawn, believing only that they are very likely not to be drawn. But if this is possible, it follows that belief simpliciter is not the same thing as degree of belief coming up to the threshold.

At this point the subjectivist normally despairs of making sense of belief simpliciter, insisting that only degrees of belief make sense.[5]

## 4. Subjective Expected Utility Theory

Subjective expected utility theory bases rational decision making on subjective probabilities. Subjectivists normally assume that the cognizer always has a subjective probability for every proposition *P*. Subjective expected utility theory then proposes that rational decisions should be made on the basis of what the rational cognizer's subjective probabilities *actually are*. If a cognizer firmly believes that *P* is true, and he does so rationally (so his degree of belief is the same as his subjective probability), then he should be willing to accept risky odds on *P*. If he believes *P* only tentatively, he should be more reluctant to bet. Rational choices are determined by how firmly the agent believes things. It is important to the subjectivist that this does not require the cognizer to have *beliefs* about his subjective probabilities. It is the actual values of his subjective probabilities, not the believed values (if he has any beliefs about that) that are relevant to decision making. This is important because, as just discussed, the subjectivist has typically denied that belief simpliciter, as opposed to degrees of belief, even makes sense, and on that view there can be no such things as beliefs about subjective probabilities. There might be varying degrees of belief to the effect that the subjective probability has different values (or lies in certain ranges of values), but it makes no sense to talk about "the subjective probability *P* is believed to have". Thus on subjective expected utility theory, the optimality prescription tells us that a rational choice is one that *in fact* maximizes expected values, not one that the agent *believes* to maximize expected values.

Subjective expected utility theory seems to absolve the agent of the epistemological task of finding out how probable things are. By contrast, objective expected utility theories have a harder time with probabilities. Because objective probabilities are facts about the world, in order to use them in decision making an agent must have beliefs about them. For such theories it seems

---

[5] Kaplan (1996) proposes an alternative account in terms of assertion rather than degrees of belief, but even he eventually concludes (p. 142) that "the Assertion View does not answer to our ordinary notion of belief. But then no coherent account of belief can."

that the optimality prescription must be interpreted as saying that a choice is rational iff the decision maker rationally *believes* that it maximizes expected value. This has two consequences. First, the decision maker must do the epistemological dirty work of really investigating the probabilities. Subjectivists sometimes get away with saying, "That is my degree of belief, and that is the end of the matter." The rest of us need reasons for our beliefs about probabilities. Thus it is incumbent on a non-subjective theory to couple the theory of practical cognition with a theory of epistemic cognition about probabilities.

The second consequence turns on the observation that, on a non-subjective expected utility theory, a decision maker may not always have beliefs about the requisite probabilities. Sometimes we are simply ignorant. In that case we cannot apply the optimality prescription. Economists distinguish between *decision making under risk* and *decision making under uncertainty* (Luce and Raiffa 1957). Decision making under risk occurs when the agent has justified beliefs about the relevant probabilities. Decision making under uncertainty occurs when he does not. In the latter case, decisions cannot be based on comparisons of expected value. Instead, game-theoretic principles like *maximin* are often proposed. However, for subjective expected utility theory there is no such thing as decision making under uncertainty. All decision making is decision making under risk.

Because I am going to come down on the side of a non-subjective expected utility theory, decision making under uncertainty is a problem for me. My theory cannot handle it. So the theory propounded in this book is only a theory of decision making under risk. This makes it an incomplete theory of rational choice. Human beings get around this to some extent by employing the shortcut procedures discussed in chapter 4, but that is not a complete solution to the problem.

## 5. Rational Decision Making

I have sketched the theory of subjective probabilities and its use in subjective expected utility theory. But now let us consider how reasonable this is as a theory of rational decision making.

Subjective expected utility theory claims not to have to deal with decision making under uncertainty, because (it is claimed) subjective probabilities always exist, and the decision maker does not have to have beliefs about their values. This is the orthodox view, but we can reasonably question whether it is correct. If the agent is to make decisions on the basis of subjective probabilities without having beliefs about them, then the subjective probabilities must somehow be directly accessible to his cognition. This is not quite as odd as it might seem at first glance. Note that on more standard views of belief, cognition has direct access to what the agent believes—the agent does not have to have beliefs about his beliefs in order to reason. The subjectivist must think that subjective probabilities play an analogous role in cognition. Is this at all plausible? An agent's *actual* degrees of belief might be accessible to his cognition in a way that does not require him to have beliefs about them. However, the subjective probability of $P$ is supposed to be the degree of the belief the agent *ideally rationally ought to have* in $P$, not necessarily his actual degree of belief. It is hard to see how cognition could have direct access to the degree of belief the agent rationally ought to have if that can be different from his actual degree of belief.

This problem has often been overlooked because subjective expected utility theory has generally been viewed as a theory of ideal rationality. This is to take it as a theory of how an ideally rational agent should make decisions, and for an ideally rational agent, the degree of belief he rationally ought to have is the same as his actual degree of belief. But in this book we are looking for a theory of rational decision making for real agents, not ideal agents.

One might attempt to avoid this problem by separating the demands of rationality. The suggestion would be that *practical* rationality requires the cognizer to make decisions based upon his actual degrees of belief, but it is *epistemic* rationality that requires the decision maker to have actual degrees of belief that are coherent. Even if the cognizer fails the test of epistemic rationality, we can still evaluate his decision making by reference to his actual degrees of belief. The objective of this separation is to enable the cognizer to avoid decision making under uncertainty by engaging in decision making without having beliefs about subjective probabilities. He may still need beliefs about probabilities in order to achieve epistemic rationality, but not for decision making.

Unfortunately, this division of rational labor cannot be made to work. The optimality prescription has an absurd consequence when expected values are computed in terms of actual degrees of belief. It implies that all bets are rational—it is impossible to bet irrationally. To see how this follows, suppose I am offered the opportunity of betting at $r{:}n$ odds that $P$ is true. My choice is between accepting the bet and not accepting the bet. The optimality prescription says that I should accept the bet iff the expected value of doing so is greater than the expected value of not doing so. But a simple calculation reveals that if we use actual degrees of belief to compute expected values, then the expected value of accepting this bet is greater than the expected value of not accepting the bet iff my degree of belief in $P$ is greater than $n/(n+r)$. However, by the definition of "degree of belief" given above, my degree of belief in $P$ is greater than $n/(n+r)$ iff I would accept a bet that $P$ is true with odds $r{:}n$. Thus what the optimality prescription tells us about this choice is that it is rational to accept the bet iff I would accept the bet. In other words, rationality imposes no constraint. It is rational to accept whatever bet I want to accept. Surely, this is wrong. The whole point of rationality is to evaluate choices and tell us that some are good choices and others are bad choices.

The upshot of these considerations is that subjective expected utility theory may not have quite so smooth sailing as people often suppose. At least when it is applied to real agents rather than ideal agents, the only way agents could conform to the optimality prescription is by having beliefs about probabilities. These are beliefs simpliciter. There must be values the probabilities are believed to have if we are to be able to compute expected values for actions and apply the optimality prescription. Thus, contrary to the orthodox claim, the subjectivist must find a way of making sense of belief simpliciter, or at least a way of accommodating its existence within his theory. Furthermore, the subjectivist will face the problem of decision making under uncertainty just as his more objective brethren do.

## 6. Do Subjective Probabilities Exist?

Subjectivists typically assume that subjective probabilities always exist. That is, for every cognizer *S* and proposition *P*, there is a unique number *p* that is *S*'s subjective probability for *P*. However, it is hard to see why this should be the case. First, it is likely that, for many propositions, a cognizer will not even have a degree of belief. If I am genuinely ignorant about *P*, my rational stance might be to refuse to bet rather than to bet in some particular way. Of course, I might be forced to bet, in which case I would acquire a degree of belief in the technical sense of being willing to accept some particular odds on a bet that *P* is true. But it seems that if I am truly ignorant, the best I can do is pick a number *p* at random. If so, then even though I may acquire a degree of belief, it seems there is no unique degree of belief that I rationally ought to have. So the subjective probability does not exist.

Against this, it might be urged that if I am truly ignorant about *P*, then I should rationally adopt a degree of belief 0.5 in *P*, in which case there is a determinate degree of belief I ought to have. However, applying this prescription generally yields degrees of belief that do not conform to the probability calculus. For example, while recognizing that *P* and *Q* are logically incompatible, I can be completely ignorant about whether *P* is true, and about whether *Q* is true, and also about whether the disjunction ($P \vee Q$) is true. For instance, let *P* be the proposition that the 327*th* automobile to pass through a certain intersection this morning is blue, and let *Q* be the proposition that it is green. I cannot consistently assign 0.5 as the subjective probabilities of all of *P*, *Q*, and ($P \vee Q$). Because *P* are *Q* are jointly inconsistent and have nonzero probabilities, the probability calculus entails that $\text{PROB}(P \vee Q) > \text{PROB}(P)$ and $\text{PROB}(P \vee Q) > \text{PROB}(Q)$. So one cannot consistently assign probabilities of 0.5 to all propositions regarding which one feels completely ignorant, but there is no obvious way to be more selective. So why should we suppose that for every cognizer *S* and proposition *P*, there is a unique number *p* that is the unique degree of belief *S* rationally ought to have in *P*?

The preceding is an argument that subjective probabilities do not *always* exist. The situation gets worse when we reflect upon the fact that even when real agents do not regard themselves as completely ignorant, and are willing to bet on propositions, their actual degrees of belief are not coherent. Given a real agent with incoherent degrees of belief, presumably the degrees of belief he rationally ought to have are a function of what degrees of belief he actually has. If there are to be unique rational degrees of belief, they must be obtained by repairing the incoherences in the agent's actual degrees of belief. But how is this to be done? If the only constraint is that the repaired degrees of belief be coherent, there will always be infinitely many ways of repairing the initial incoherent set of degrees of belief, and for any contingent proposition $P$ and number $p$ between 0 and 1, one of the resulting coherent sets of degrees of belief will assign $p$ as the probability of $P$. Even if we impose plausible constraints like minimally changing the actual degrees of belief to render them coherent, there is no reason to expect there to be a unique way of doing that, and so no reason to expect contingent propositions to have well-defined rational degrees of belief.

Might there be additional constraints on degrees of belief that would yield unique rational degrees of belief to serve as subjective probabilities? Subjectivists sometimes explicitly assert that the only rationality constraint is conformance to the probability calculus. Some subjectivists will add a principle of Bayesian updating and insist that it together with coherence constitute the only rationality constraints (Howson and Urbach 1989). Some follow David Lewis (1980, 1994) in requiring that the "principal principle" be satisfied. But the number of additional constraints that have been proposed is small, and they do not come anywhere close to generating unique rational degrees of belief.[6] If this is all there is to rational degrees of belief, then it must be concluded that subjective probability, defined as the unique degree of belief that a real agent rationally ought to have in a proposition, simply does not exist. So defined, subjective probability is a myth. The concept of the subjective probability of a proposition $P$ for a real agent is simply unintelligible.

Perhaps for this reason, subjectivists like David Lewis and Brian Skyrms (1980, 1984) tend to talk about subjective probabilities only in connection with ideally rational agents. The Dutch book argument may convince us that if an agent is ideally rational he will have coherent degrees of belief. In that case, for ideally rational agents, subjective probabilities can be identified with actual degrees of belief. As long as we suppose that ideally rational agents will have degrees of belief regarding all propositions, it follows unproblematically that they will have subjective probabilities conforming to the probability calculus. Then we might plausibly claim that ideally rational agents should base their decisions on comparisons of expected value, where the latter are computed using their ideally rational degrees of belief. However, even if this were true, it is hard to see what this has to do with rational decision making by real cognizers, for whom subjective probabilities do not exist. As we saw in chapter 1, theories of ideal rationality tell us little about real rationality. If subjective expected utility theory is a theory of ideal rationality, then it leaves unanswered the question how real decision makers should make decisions, and that is the question this book aims to answer.

---

[6] Kaplan (1996) claims there are more constraints, but he does not attempt to tell us what they are. However, his theory is of a different sort (discussed in the next section) and is not committed to there being unique rational degrees of belief.

## 7. Deriving the Optimality Prescription from Rationality Constraints

There is another strand of subjective expected utility theory that might seem to avoid the preceding difficulties. It was remarked in chapter 2 that much work has gone into deriving versions of the optimality prescription from what are proposed as rationality constraints on either preferences alone or preferences and degrees of belief together. The idea comes originally from Ramsey (1926). Variants of it can be found in von Neumann and Morgenstern (1944), Savage (1954), Bolker (1966), Jeffrey (1983), Armendt (1986, 1988), Kaplan (1996), and Joyce (1998). The strategy is to argue that our preferences and degrees of belief should, rationally, satisfy certain conditions. For instance, most authors agree that preferences should be transitive. Not all of the rationality constraints proposed by these authors are as intuitively compelling as this, but let us waive that difficulty for the moment. It is then shown that if our degrees of belief and preferences conform to the constraints, then we will make decisions in accordance with some version of the optimality prescription.

Thus far, what we have is simply a mathematical theorem. But the attempt is made to turn it into an argument that we should, rationally, make decisions in accordance with some version of the optimality prescription.[7] The general form of the argument is that we should, rationally, have degrees of belief and preferences that conform to the rationality constraints; but in order to do that we must make decisions in accordance with the optimality prescription; so we should, rationally, make decisions in accordance with the optimality prescription.

What is novel about this argument is that no claim is made that there are unique degrees of belief and preferences that we rationally ought to have. The constraints on preferences and degrees of belief are purely formal constraints. If this form of argument works, it promises to get us the optimality prescription without our having to make sense of subjective probabilities. This is because the form of the optimality prescription that is derived from this argument appeals only to actual degrees of belief. The premise of the argument is that our actual degrees of belief ought to satisfy certain constraints, and the conclusion is that we ought to make decisions based on the optimality prescription as applied to our actual degrees of belief. Thus we avoid all the difficulties encountered in trying to make sense of subjective probabilities.

Unfortunately, the argument is fallacious. It has the form "I ought to do *A*; doing *A* requires doing *B*; so I ought to do *B*." Consider an instance of this argument. Suppose that, for some reason, I ought to fly to Cleveland. That requires that I purchase an airline ticket for a flight to Cleveland, so the conclusion is that I ought to purchase such a ticket. This might seem

---

[7] Kaplan (1996) is perhaps the clearest example of this, although the version of the optimality prescription that he endorses is weaker than the standard version.

right. But now, suppose that I decide not to fly to Cleveland, even though I should. Is it still true that I ought to purchase an airline ticket for a flight to Cleveland? Surely not. The difficulty is that "I ought to do *A*" is ambiguous. It might mean, "In any world in which all obligations were satisfied, I would do *A*", or it could mean "In the actual world, with all of its failings, I ought to do *A*." In the former sense, I ought to buy a ticket, but not the in latter sense.

We can apply this same distinction to the argument for the optimality prescription. Assuming the correctness of the proposed rationality constraints, it would be true that in any world in which all rationality constraints are satisfied, I will conform to the optimality prescription (applied to actual degrees of belief). But it does not follow that in the real world, in which I may not conform to the rationality constraints, I should conform to the optimality prescription (applied to actual degrees of belief). In fact, we have already seen that this conclusion is incorrect. We noted in section 5 that if we endorse the optimality prescription interpreted in terms of our actual degrees of belief, it has the absurd consequence that we ought to accept whatever bets we do accept, that is, irrational betting is impossible. So this way of defending the optimality prescription does not work. It runs afoul of the same difficulties as the more standard versions of subjective expected utility theory. That is, it only allows us to conclude that an agent should conform to the optimality prescription if the agent conforms to all of the rationality constraints. In other words, it only applies to ideal agents.

## 8. Subjective Probabilities from Epistemology

A defensible version of the optimality prescription must proceed in terms of something other than actual degrees of belief. The subjectivist must find some way of understanding subjective probabilities that makes them well defined but different from actual degrees of belief. We might attempt to do this by turning to yet another strand of the theory of subjective probabilities. Subjective probabilities have been used for two purposes. Thus far we have concentrated on their use in practical decision making, but it has also been claimed that they can provide the basis for an epistemological theory of rational belief—*Bayesian epistemology*. Bayesian epistemology has not been popular among mainstream epistemologists, but it has had its influential proponents, for example, David Lewis (1980) and Brian Skyrms (1980, 1984). A version of it is worked out in detail in Howson and Urbach (1989).

The difficulties I have raised for subjective probability as ideally rational degree of belief apply just as much to its use in epistemology as to its use in rational decision making. However, it might be suggested that a different way of making sense of subjective probability is to turn Bayesian epistemology on its head. Instead of explaining epistemic justification in terms of subjective probabilities, perhaps we can adopt a more conventional epistemological account of epistemic justification and then identify the subjective probability of a proposition with its degree of justification. The idea is that the degree of belief an agent ought to have in a proposition is determined by the degree to which he is justified in believing it. This would yield a decidedly non-conventional subjective expected utility theory, but it may be one way of making sense of subjective probabilities.

### *8.1 Do Degrees of Justification Conform to the Probability Calculus?*

This way of making sense of subjective probabilities will only work if degrees of justification conform to the probability calculus. Unfortunately, there is a simple argument that establishes conclusively that they do not. Recall that we axiomatized the probability calculus as follows:

THE PROBABILITY CALCULUS:

(1) **PROB**($P$ & ~$P$) = 0.
(2) **PROB**($P \vee$ ~$P$) = 1.
(3) **PROB**($P \vee Q$) = **PROB**(P) + **PROB**($Q$) − **PROB**($P$ & $Q$).
(4) If $P$ and $Q$ are logically equivalent then **PROB**($P$) = **PROB**($Q$).

If $Q$ is a necessary truth, it is logically equivalent to ($P \vee$ ~$P$), so it follows from axioms (2) and (4) that every necessary truth has probability 1. If degrees of justification satisfied the probability calculus, mathematics would be terribly easy. As necessary truths, mathematical truths automatically have probability 1. If that meant that everyone is automatically justified in believing all mathematical truths, then we would have no need for mathematical proofs. The absurdity of this conclusion indicates that degrees of justification do not conform to the probability calculus.

It might be supposed that the culprit here is axiom (4). In mathematical probability theory, probabilities are supposed to apply to "events", not propositions. Axiom (4) was motivated by the suggestion that events, in the relevant sense, can be identified with sets of logically equivalent propositions. But perhaps that was too quick. Can we avoid the problem by deleting axiom (4)? It turns out that we cannot. It can be shown to follow from just axioms (1)–(3) that every tautology (of the propositional calculus) has probability 1. For example, although this is not obvious on casual inspection, $[P \leftrightarrow (Q \,\&\, {\sim}P)] \rightarrow {\sim}Q$ is a tautology. It follows from axioms (1)–(3) that it has probability 1. Again, if degrees of justification satisfied the probability calculus it would follow that everyone is automatically justified in believing any proposition having this complex form, even before they have seen that it is a tautology. This is no less absurd than the analogous principle regarding mathematical truths.

### *8.2 Do Degrees of Warrant Conform to the Probability Calculus?*

In chapter 1 I distinguished between warrant and justification. Warranted propositions are propositions that an agent would be justified in believing if, starting from his current epistemic state, he could complete all the relevant reasoning. Although tautologies are not automatically justified, they are automatically warranted. They can always be proven if the agent does enough reasoning. This suggests that we might identify subjective probabilities with degrees of warrant rather than degrees of justification. That avoids the problem with tautologies. It is less clear whether it will salvage axiom (4). If, perhaps on the grounds of Gödel's theorem, one thinks there could be unprovable necessary truths, then not all necessary truths will be warranted. An unprovable necessary truth is still logically equivalent (but not provably equivalent) to $(P \vee \sim P)$, so axiom (4) would fail. We might, however, replace axiom (4) with a weaker axiom appealing to provable equivalences.

### *8.2.1 Probabilism and Reasoning*

Although tautologies are always warranted, this does not yet establish that degrees of warrant conform to the probability calculus. Let us call the view that they do *probabilism*. There are, in fact, familiar arguments that purport to show that probabilism is false. The simplest has to do with the role of reasoning in epistemology. Although different epistemological theories may give differing accounts of reasoning, they are virtually unanimous in agreeing that reasoning plays an essential role in our knowledge of the world. Some of our knowledge is acquired directly from perception, but without reasoning all perception can tell us is how our immediate environment looks and feels. Most of our knowledge is acquired indirectly by reasoning in various ways from our perceptual knowledge. The whole point of reasoning is that it enables us to become justified in holding new beliefs on the basis of other beliefs we are already justified in holding. The simplest argument against probabilism is that it would make it impossible for a conclusion to be warranted on the basis of a deductive argument from multiple uncertain premises. This is because, if degrees of warrant work like probabilities, then when you combine premises, the degree of warrant decreases. Suppose you have 100 independent premises, each with a degree of warrant of .99. If probabilism is true, then by the probability calculus, the degree of warrant of the conjunction will be only .37, so we could never be warranted in using these 100 premises conjointly in drawing a conclusion. But this flies in the face of common sense. For example, consider an opinion pollster surveying people about which of two products they prefer. She surveys 100 people, collecting from each the verbal expression of an opinion of the form "I prefer $x$ to $y$". She summarizes her data by saying, "I surveyed 100 people, and 79 of them reported preferring $A$ to $B$." This conclusion follows deductively from her accumulated data. But each piece of data of the form "Person $S$ reported preferring $x$ to $y$" is something she believes with less than certainty—we are supposing she believes it with a degree of warrant of .99. Then if degrees of warrant work like probabilities, her degree of warrant for thinking that she has surveyed 100 people and 79 of them reported preferring $A$ to $B$ would be only .37, and hence she would not be warranted in drawing that conclusion. Surely this is wrong.

Consider another example—counting apples in a barrel. Let us suppose you a very meticulous counter. You examine each apple carefully as you remove it from the barrel, ascertaining that it is indeed an apple, and you then carefully jot down a mark to count the apple, using horizontal and vertical marks like  so that when you are finished you can read off the result as a number. Let us suppose you are virtually certain you have not lost count (your degree of warrant in that is .999), so the only source of uncertainty is in your judgments that the individual objects counted are apples. Suppose you count $n$ apples, judging each to be an apple with a degree of warrant $j$. If degrees of warrant work like probabilities, the probability calculus reveals that your degree of warrant for believing that there are at least $r$ ($r \leq n$) apples in the barrel will be

$$\sum_{i=r}^{i=n} j^{r}(1-r)^{n-i} \frac{n!}{r!(n-i)!}$$

So, for example, if you count 100 apples in the barrel, being warranted to degree .95 in believing that each object counted is an apple, then your degree of warrant for believing that there are 100 apples in the barrel is only .006. Your degree of warrant for believing that there are at least 96 apples in the barrel is only .258. You have to drop all the way down to the judgment that there are at least 95 apples in the barrel before you get a degree of warrant greater than .5. If you want a degree of warrant of at least .95 for your judgment of the number of apples in the barrel, the best you can do is conclude that there are at least 91. So on this account, you cannot even count apples in a barrel. Similarly, if you have 6 daughters, and your degree of warrant for believing of each that she is indeed one of your daughters is .95, then all you can be justified in believing to degree .95 is that you have at least 5 daughters. Surely this is ridiculous.

### 8.2.2 *Conjunctivitis*

Still, there are philosophers (for example, Kyburg 1970) who have been willing to bite the bullet and deny that deductive reasoning from warranted premises conveys warrant to the conclusion. We can make a distinction between two kinds of deductive inference rules. Let us say that a rule is *probabilistically valid* iff it follows from the probability calculus that the conclusion is at least as probable as the least probable premise. For instance, *simplfication* and *addition* are probabilistically valid:

*Simplification*: $(P \& Q) \vdash P$

*Addition*: $P \vdash (P \vee Q)$.

But not all familiar inference rules are probabilistically valid. For example, it is widely recognized that *adjunction* is not:

*Adjunction*: $P,Q \vdash (P \& Q)$.

In general, PROB($P$ & $Q$) can have any value between 0 and the minimum of PROB($P$) and PROB($Q$). Because of this, Kyburg claims that it is a fallacy to reason using adjunction. He calls this fallacy "conjunctivitis". For those who are persuaded by these considerations, the view would be that we are only allowed to reason "blindly", without explicitly computing probabilities (or degrees of warrant) as we go along, when the rules of inference we use are probabilistically valid. In all other cases, we must compute the probability of the conclusion to verify that it is still sufficiently probable to be believable. Probabilism is committed to this view. If degrees of warrant satisfy the probability calculus, then without computing probabilities we can only be confident that a deductive argument takes us from warranted premises to a warranted conclusion if all of the inferences are probabilistically valid.

Which deductive inference rules are probabilistically valid? It is easily shown that any valid deductive inference rule proceeding from a single premise is probabilistically valid. On the other hand, some rules proceeding from multiple premises are not. For example, *adjunction* is not. Are there others? People are generally amazed to discover that *no* deductive inference rule that proceeds from multiple premises essentially (that is not still valid if you delete an unnecessary premise) is probabilistically valid. They all go the way of adjunction. For instance, *modus ponens* and *modus tollens* are not probabilistically valid. Probabilistic validity is the exception rather than the rule.

The upshot of this is that if probabilism is true, there will be hardly any deductive reasoning from warranted premises that we can do blindly and still be confident that our conclusions are warranted. Blind deductive reasoning can play very little role in epistemic cognition. Epistemic cognition must instead take the degrees of justification of the premises of an inference and compute a new degree of justification for the conclusion in accordance with the probability calculus.

This might not seem so bad until we realize that it is impossible to do. The difficulty is that the probability calculus does not really enable us to compute most probabilities. In general, all the probability calculus does is impose upper and lower bounds on probabilities. For instance, given degrees of justification for $P$ and $Q$, there is no way we can compute a degree of justification for $(P \& Q)$ just on the basis of the probability calculus. As just noted, given probabilism, it is consistent with the probability calculus for the degree of justification of $(P \& Q)$ to be anything from 0 to the minimum of the degrees of justification of $P$ and $Q$ individually. Another way of looking at this is to note that, by the probability calculus, $\textbf{PROB}(P \& Q) = \textbf{PROB}(Q)\cdot\textbf{PROB}(P/Q)$. If degrees of justification work like probabilities, then we can define a conditional degree of justification $\textbf{DJ}(P/Q)$ such that $\textbf{DJ}(P \& Q) = \textbf{DJ}(Q)\cdot\textbf{DJ}(P/Q)$. If $P$ and $Q$ are independent then $\textbf{DJ}(P/Q) = \textbf{DJ}(P)$, but if $Q$ is negatively relevant to $P$ then $\textbf{DJ}(P/Q)$ can range from 0 to $\textbf{DJ}(P)$, and if $Q$ is positively relevant to $P$ then $\textbf{DJ}(P/Q)$ can range from $\textbf{DJ}(P)$ to 1. There is in general no way to compute $\textbf{DJ}(P/Q)$ just on the basis of logical form. The value of $\textbf{DJ}(P/Q)$ is normally a substantive fact about $P$ and $Q$, and it must be obtained by some method other than mathematical computation in the probability calculus.

These observations lead to a general, and I think insurmountable, difficulty for probabilism. Probabilism claims we must compute degrees of justification as we go along in order to decide whether to accept the conclusions of our reasoning. If conditional degrees of justification conform to the probability calculus, they will generally be idiosyncratic, depending upon the particular propositions involved. That is, they cannot be computed from anything else. If they cannot be computed, they must be stored innately. This, however, creates a combinatorial nightmare analogous to the problems we encountered in chapter 2 for stored preferences. As Gilbert Harman (1973) observed years ago, given a set of 300 propositions, the number of probabilities of single propositions conditional on conjunctions of propositions in the set is $2^{300}$, which is greater than a recent estimate of the number of elementary particles in the universe. It is computationally impossible for a real agent to store that many primitive probabilities.[8] Thus probabilism would make reasoning impossible.

The upshot of this is that if probabilism were true, we could not acquire new justified beliefs by reasoning from previously justified beliefs. However, reasoning is an essential part of epistemic cognition. Without reasoning, all we could know is that our current environment looks and feels various ways to us. It is reasoning that allows us to extend this very impoverished perceptual knowledge to a coherent picture of the world. So probabilism cannot be right.

### *8.3 Conclusions*

If neither degrees of justification nor degrees of warrant conform to the probability calculus, then we cannot make sense of subjective probability by identifying it with one of these candidates. So we are still left without a way of making sense of subjective probabilities.

## 9. A Return to Objective Probabilities

Historically, theories of objective probability predated theories of subjective probability. However, what were perceived as fatal difficulties for existing theories of objective probability led philosophers to look for an alternative. The alternative they found was subjective probability. But I have now argued that standard approaches to subjective probability theory fare no better than the classical theories of objective probability. They do not succeed in making subjective probabilities intelligible. This suggests that we should revisit objective probabilities and see whether they are quite so badly off as they are generally supposed to be. This will be undertaken in the next chapter, where it will be argued that the standard objections to objective probabilities were based on theories of philosophical analysis that have

[8] Of course, we might not be necessary to store them all. We might, for example, use a Bayesian net to omit all those cases in which the propositions are statistically independent. However, it is easy to construct cases in which every proposition is statistically dependent on every conjunction of other propositions in the set. An example is the "slippery blocks" problem described in chapter 10.

long since been discredited. So an attempt will be made to resurrect objective probabilities and investigate the extent to which they can be employed in rational decision making.

Are we being too quick to reject subjective probability? For example, Kaplan (1996) repeatedly observes that, surely, there is some sense of "degree of confidence" for which principles like the following seem to be correct:

> If you are more confident that *P* will be true than that ~*P* will be true, and you are offered a bet which will pay you $10 if *P* turns out to be true but will require you to pay $1 if *P* turns out to be false, then, other things being equal, it would be rational to accept this bet.

Such degrees of confidence *sound like* the material for building theories of subjective probability. However, once we have decided to give objective probabilities another look, there is an obvious suggestion regarding how to understand these degrees of confidence. They are just the believed values for the objective probabilities. You are more confident that *P* will be true than that ~*P* will be true just in case you think *P* is objectively more probable than ~*P*. That will be the suggestion of chapter 7.

# 7

# Objective Probabilities

In the last chapter, I argued that there is no way to make sense of subjective probability, and I urged a return to objective probabilities. This chapter aims at making sense of objective probability and rebutting the arguments that led, historically, to its falling into disrepute.

Philosophers often fail to realize that there are a number of different kinds of probability, with different mathematical and epistemological properties. In this chapter I will first introduce the reader to nomic probability, which is the kind of probability involved in probabilistic laws of nature. I will sketch the theory of nomic probability introduced in my book *Nomic Probability and the Foundations of Induction* (Pollock 1990). A more detailed sketch is presented in the appendix, but for the full details the reader should consult the book.

Nomic probabilities themselves are not plausible candidates for use in decision-theoretic reasoning. However, it will be shown in section 4 that by appealing to nomic probabilities we can define a kind of "mixed physical/epistemic probability" that is plausibly of use in decision making. It will turn out, in chapter 8, that mixed physical/epistemic probabilities are still not quite the right probabilities for use in defining expected values, but it will be shown how to use them to define a concept of "causal probability" which, I believe, does the job.

## 1. Physical Probabilities and Relative Frequencies

Most historical theories of objective probability have been theories of "physical probability" and they proceeded by relating probability to frequency. Physical probabilities are supposed to be logically contingent physical features of the world. The relative frequency of $A$'s in $B$'s, symbolized "freq[$Ax/Bx$]", is the proportion of all $B$'s that are $A$'s. If there are $m$ $B$'s and $n$ of them are $A$'s, then freq[$Ax/Bx$] = $n/m$. Note that frequencies relate properties, which we symbolize using formulas containing free variables. We can talk about relative frequencies either in a small sample of $B$'s, or relative frequencies among all the $B$'s in the world. Many theorists were tempted by the idea that probability could be defined in terms of relative frequency. The simplest such theory identifies probabilities with relative frequencies. Among the modern proponents of this theory are Bertrand Russell (1948), R. B. Braithwaite (1953), Henry Kyburg (1961, 1974), and

Lawrence Sklar (1970, 1973).[1]

Theories identifying probability with relative frequency are generally considered inadequate on the grounds that we often want to talk about probabilities even in cases in which relative frequencies do not exist. In this connection, note that freq[$Ax/Bx$] does not exist if either there are no *B*'s or there are infinitely many. For example, applying this to quantum mechanics, a certain configuration of particles may never have occurred, but we may still want to talk about the probability of the emission of a photon from such a configuration. Quantum mechanics even tells us how to calculate such probabilities. But as such a configuration has never occurred, there is no relative frequency with which the probability can be identified.

A related difficulty occurs when the reference class is infinite. The relative frequency does not exist in this case either. It is apt to seem that the infinite case is less problematic because we are rarely faced with infinite totalities. But that is a mistake. In talking about the probability of a *B* being an *A*, we are not usually referring to the probability of a *B* being an *A now*. Unless we explicitly relativize the properties to times and ask for the probability of a *B at time t* being an *A*, then we are typically asking about the probability at an arbitrary time. The relative frequency would then be the proportion of *B*'s that are *A*'s throughout all time, and in most cases there is no guarantee that there will be only finitely many *B*'s throughout the history of the universe. For example, unless we adopt certain cosmological theories, there is no compelling reason to think that there will be only finitely many stars throughout the history of the universe, or finitely many electrons, or even finitely many people. But then an identification of probabilities with relative frequencies will leave most probabilities undefined.

Theories identifying probabilities with actual relative frequencies in finite sets have come to be called *finite frequency theories*. In order to avoid the problem of the infinite reference class, most proponents of frequency theories have turned instead to *limiting frequency theories*. These theories identify probabilities with *limits of relative frequencies* rather than with relative frequencies simpliciter.[2] The idea is that if there are infinitely many *B*'s, then the probability of a *B* being an *A* is defined as the limit the relative frequency approaches as we consider more and more *B*'s. In the special case in which there are only finitely many *B*'s, the probability is identified with the actual relative frequency. It is now generally recognized that limiting frequency theories face extreme difficulties. First, they are of no help at all in the case in which there are no *B*'s. What is initially more surprising is that they do not work in the infinite case either. The difficulty is that the notion of the

---

[1] William Kneale (1949) traces the frequency theory to R. L. Ellis, writing in the 1840s, John Venn (1888), and C. S. Peirce in the 1880s and 1890s. It is slightly inaccurate to include Kyburg in this camp, because he does not use the term "probability" for relative frequencies, reserving it instead for definite probabilities (see following), but what plays the role of indefinite probability in his theory is relative frequency.

[2] This suggestion can be traced to John Venn (1888), p. 95. It has been defended in detail by von Mises (1957), Popper (1959), and Reichenbach (1949). Kyburg (1974) makes passing concessions to limiting frequency theories, but the spirit of his theory is finite frequentist.

limit of relative frequencies of *A*'s in *B*'s makes no mathematical sense. If there are infinitely many *B*'s, and infinitely many of them are *A*'s and infinitely many of them are non-*A*'s, then you can obtain whatever limit you want by going through the *B*'s in different orders. We might begin by enumerating the *B*'s in some order $b_1,...,b_n,...$ . Then let us construct a new ordering on the basis of the first. We begin by picking the first *B* that is an *A*. Then we pick the first *B* that is not an *A*. Then we pick the next two *B*'s that are *A*'s, followed by the next *B* that is not an *A*. Then we pick the next four *B*'s that are *A*'s, followed by the next *B* that is not an *A*; and so on. The limit of the relative frequency obtained in this way will be 1. If instead we reversed the above procedure and first picked *B*'s that are not *A*'s, we would arrive at a limit of relative frequencies of 0. And we can construct other orderings that will produce any limit between 0 and 1. The point is that there is no such thing as "the limit of relative frequencies of *A*'s in *B*'s". It only makes sense to talk about the limit relative to a particular ordering of the *B*'s. Different orderings yield different limits.

For a while it was hoped that limiting frequency theories could be salvaged by placing some natural constraints on the ordering of the *B*'s, and much effort went into the search for such constraints. The original idea goes back to von Mises (1928, 1957). He proposed to define the probability of a *B* being an *A* in terms of *collectives*, where a collective is any ordering of the *B*'s that is "random" in the sense that the same limiting frequency is exhibited by any infinite subsequence drawn from the whole sequence by "place selections". A place selection is an algorithm for choosing elements of the whole sequence in terms of their place in the sequence. If no restrictions are imposed on place selections then there are always place selections that generate subsequences with different limits.[3] Alonzo Church (1940) proposed to solve this problem by requiring place selections to be recursive functions, and Church's proposal was subsequently refined by other authors.[4] However, this whole approach seems misguided. Invariance over changes in place selections does not seem to have anything to do with the relationship between prob(*Ax*/*Bx*) and the behavior of sequences of *B*'s. Suppose, for example, that it is a physical law that *B*'s occur sequentially, and every other *B* is an *A*. An example might be the temporal sequence of days and nights. This sequence is completely nonrandom and violates any criterion formulated in terms of place selections, but we would not hesitate to judge on the basis of this sequence that prob(*Ax*/*Bx*) = 1/2.

It is now generally acknowledged that the search for constraints on sequences has been in vain, and accordingly the limiting frequency theories have lost much of their popularity.[5]

---

[3] See Salmon (1977) for a discussion of the history of the development of this concept.

[4] See particularly Martin-Löf (1966, 1969) and Schnorr (1971).

[5] For critical discussions of limiting frequency theories along this and other lines, see Russell (1948) (pp. 362ff), Braithwaite (1953; p. 125), and Kneale (1949; pp. 152ff).

## 2. Empirical Theories

It is simplistic to identify physical probabilities with either relative frequencies or limits of relative frequencies. Nevertheless, it appears that there must be important connections between physical probabilities and relative frequencies. At the very least there must be epistemic connections. It is undeniable that many of our presystematic probability judgments are based upon observations of relative frequencies. For example, the meteorologist assesses the probability of rain by considering how often it has rained under similar circumstances in the past. I will call any theory of physical probability that draws a tight connection between probabilities and relative frequencies an *empirical theory*. In constructing a theory of objective probability, our task is to defend some kind of empirical theory.

Theories of subjective probability were born in the first half of the twentieth century (Ramsey 1926, Savage 1954) in response to the perceived failure of frequency-based theories of objective probability. Indeed, all theories of objective probability then in existence were failures. But it is too quick to generalize from this to the conclusion that the very concept of objective probability is philosophically suspect. Human beings reason about objective probabilities all the time. It is typical of philosophically interesting concepts that we know how to employ them in reasoning without knowing how we do it. Discovering how we do it is the task of the epistemologist.

The mid-twentieth century controversy about objective probability occurred in the context of an assumption that has since been completely discredited. It was assumed that the way to solve philosophical problems is to find definitions (necessary and sufficient conditions) illuminating the concepts involved. Twentieth century analytic philosophy is largely a history of the failure of this assumption. You cannot solve the problem of perception by defining "red" in terms of "looks red", you cannot solve the problem of other minds by defining "person" in terms of behavior, and you cannot provide foundations for probabilistic reasoning by defining "probability" in terms of relative frequencies. The fact that there is an epistemological connection between relative frequencies and objective probabilities is no reason at all to suppose that probabilities can be defined in terms of frequencies. You can rarely solve epistemological problems by finding definitions of the concepts involved, and the failure to find such definitions has no implications at all for the intelligibility of the concepts. Few philosophers nowadays would suppose otherwise, and yet the condemnation of objective probability on that ground seems to have stuck to it by a kind of intellectual inertia. The argument has been thoroughly discredited, but many philosophers retain the conclusion. I can only hope that by waving this observation under their noses I can get them to take a more intellectually respectable stance.

Generally, the only kind of philosophical analysis that can be given of a philosophically interesting concept is an epistemological one. We throw light on the concept by clarifying how we use it in reasoning—by explaining its epistemological place in our framework of concepts. Most interesting principles of reasoning are defeasible rather than deductive (Pollock 1986,

1995, 1998a; Pollock and Cruz 1999). For instance, we judge the color of an object by appealing defeasibly to how the object looks, and we judge that something is a person by appealing defeasibly to the way it looks and behaves. I propose that we should adopt a similar stance with regard to objective probability. The way to clarify the concept is to provide an account of how to reason about objective probabilities. What is needed is an epistemological theory of probabilistic reasoning. The principles involved will most likely be defeasible. But you cannot get defeasible principles out of a definition of "probability". Definitions only yield deductive connections. The defeasible principles themselves will play a fundamental role in characterizing the concept of probability.

Empirical theories as a group face two other difficulties. First, there is the problem of how to evaluate the probabilities. How can you discover, for example, that the probability of its raining under such-and-such circumstances is .3? It might seem that this problem would have a simple solution if we could identify probabilities with relative frequencies, but that is a mistake. We would still be faced with the problem of discovering the relative frequency. If the *B*'s are widely extended in either space or time, there can be no question of actually counting the *B*'s and seeing what proportion of them are *A*'s. Instead, we use sampling procedures. We examine a sample of *B*'s and see what the relative frequency of *A*'s is in the sample, and then we project that relative frequency (or a probability with that value) onto the class of all *B*'s. This is a form of statistical induction. A workable empirical theory of probability must provide us with some account of statistical induction. Objective probabilities are useless if there is no way to discover their values.

The second difficulty faced by empirical theories concerns a distinction between two kinds of probability. The probabilities at which we arrive by statistical induction have the same logical form as relative frequencies. That is, they concern classes of objects or properties of objects rather than specific objects. We discover the probability of *a* smoker getting lung cancer, or the probability of its raining when we are confronted with a low pressure system of a certain sort. I will call such probabilities *indefinite probabilities*.[6] It is natural to employ free variables in the formulation of indefinite probabilities. For example, we can write prob($x$ gets lung cancer/$x$ is a smoker). To be contrasted with indefinite probabilities are the probabilities that particular propositions are true or particular states of affairs obtain. For example, we can talk about the probability that Jones will get cancer, or the probability that it will rain tomorrow. These are sometimes called "single case probabilities", but I prefer to call them *definite probabilities*.[7]

---

[6] I take this term from Jackson and Pargetter (1973).

[7] The distinction between definite and indefinite probabilities is a fundamental one and, I should think, an obvious one. It is amazing how often philosophers have overlooked it and become thoroughly muddled as a result. For example, virtually the entire literature on probabilistic causation conflates definite and indefinite probabilities. See the discussion of this in chapter eight.

Empirical theories of probability are most directly theories of indefinite probabilities. This is because relative frequencies relate classes or properties, and the probabilities inferred from them must have the same logical form. But the probabilities we need for computing expected values are definite probabilities. We must know the probability that a particular outcome will occur. No theory of probability can rest content with just giving us an account of indefinite probabilities. It must also tell us how to compute the definite probabilities we need for decision making. Typically, theories of indefinite probability propose that definite probabilities can be inferred from indefinite probabilities by a species of reasoning called "direct inference". I will talk more about direct inference in section 4.

## 3. Nomic Probability

There are two kinds of physical laws—statistical and non-statistical. Statistical laws are probabilistic. I will call the kind of probability involved in statistical laws *nomic probability*. The best way to understand nomic probability is by looking first at non-statistical laws. What distinguishes such laws from material generalizations of the form "$(\forall x)(Fx \rightarrow Gx)$" is that they are not just about actual *F*'s. They are about "all the *F*'s there could be", or more precisely, they are about "physically possible *F*'s". I call non-statistical laws *nomic generalizations*. Nomic generalizations can be expressed in English using the locution "Any *F* would be a *G*".

I propose that we think of nomic probabilities as analogous to nomic generalizations. Just as the nomic generalization tells us that any physically possible *F* would be *G*, we can think of the statistical law "$\mathbf{prob}(Gx/Fx) = r$" as telling us that the proportion of physically possible *F*'s that would be *G*'s is *r*. For instance, pretend it is a law of nature that at any given time, there are exactly as many electrons as protons. Then in every physically possible world (world in which all the physical laws hold), the proportion of electrons-or-protons that are electrons is 1/2. This makes it reasonable to regard the probability of a particle being an electron given that it is either an electron or a proton as 1/2. Of course, in the general case, the proportion of *G*'s that are *F*'s will vary from one possible world to another. The probability $\mathbf{prob}(Fx/Gx)$ then "averages" these proportions across all physically possible worlds. The mathematics of this averaging process is complex, and I will not go into detail about it here. I will give a brief sketch of the general theory of nomic probability in this chapter. The theory is summarized in more detail in the appendix, and is worked out in detail in my (1990).

Nomic probability is illustrated by any of a number of examples that are difficult for frequency theories. For instance, consider a physical description *D* of a coin, and suppose there is just one coin of that description and it is never flipped. On the basis of the description *D* together with our knowledge of physics we might conclude that a coin of this description is a fair coin, and hence the probability of a flip of a coin of description *D* landing heads is 1/2. In saying this we are not talking about relative frequencies—as there are no flips of coins of description *D*, the relative frequency does not exist. Or suppose instead that the single coin of description *D* is flipped just once, landing heads, and then destroyed. In that case the relative frequency is 1, but we would still insist that the probability of a coin of that description landing heads is 1/2. The reason for the difference between the relative frequency and the probability is that the probability statement is in some sense subjunctive or counterfactual. It is not just about actual flips, but about possible flips as well. In saying that the probability is 1/2, we are saying that out of all physically possible flips of coins of description *D*, 1/2 of them would land heads.

The theory of nomic probability is a theory of probabilistic reasoning. It does not attempt to *define* "nomic probability" in terms of simpler concepts. Instead, it is an epistemological theory of how to reason about nomic probabilities. Probabilistic reasoning has three constituents. First, there must be rules prescribing how to ascertain the numerical values of nomic probabilities on the basis of observed relative frequencies. Second, there must be computational principles that enable us to infer the values of some nomic probabilities from others. Finally, there must be principles enabling us to use nomic probabilities to draw conclusions about other matters.

The first element of this account consists largely of a theory of statistical induction. The second element will consist of a calculus of nomic probabilities. The final element is an account of how conclusions not about nomic probabilities can be inferred from premises about nomic probability. This has two parts. First, it seems clear that under some circumstances, knowing that certain probabilities are high can justify us in holding related non-probabilistic beliefs. For example, I know it to be highly probable that the date appearing on a newspaper is the correct date of its publication. (I do not know that this is always the case—typographical errors do occur.) On this basis, I can arrive at a justified belief regarding today's date. The epistemic rules describing when high probability can justify belief are called *acceptance rules*. The acceptance rules endorsed by the theory of nomic probability constitute the principal novelty of that theory. The other fundamental principles that are adopted as primitive assumptions about nomic probability are all of a computational nature. They concern the logical and mathematical structure of nomic probability, and amount to nothing more than an elaboration of the standard probability calculus. It is the acceptance rules that give the theory its unique flavor and comprise the main epistemological machinery making it run.

The other kind of inference from nomic probabilities that it is important to be able to make is direct inference to definite probabilities. A satisfactory theory of nomic probability must include an account of direct inference.

To summarize, the theory of nomic probability will consist of (1) a theory of statistical induction, (2) an account of the computational principles allowing some probabilities to be derived from others, (3) an account of acceptance rules, and (4) a theory of direct inference. This section closes with a brief sketch of the theory, but readers not interested in the details can skip it and go on to section 4.

### 3.1 *The Calculus of Nomic Probabilities*

It might seem that the calculus of nomic probabilities should be the classical probability calculus. But this overlooks the fact that nomic probabilities are indefinite probabilities. Indefinite probabilities operate on properties and relations. This introduces logical relationships into the theory of nomic probability that are ignored in the classical probability calculus. One simple example is the "principle of individuals":

(IND) $\mathbf{prob}(Axy\ /\ Rxy\ \&\ y{=}b) = \mathbf{prob}(Axb/Rxb)$.

This is an essentially relational principle and is not even well-formed in the classical probability calculus. It might be wondered how there can be general truths regarding nomic probability that are not theorems of the classical probability calculus. The explanation is that, historically, the probability calculus was devised with definite probabilities in mind. The standard versions of the probability calculus originate with Kolmogorov (1933) and are concerned with definite probabilities. The relationship between the calculus of indefinite probabilities and the calculus of definite probabilities is a bit like the relationship between the predicate calculus and the propositional calculus. Specifically, there are principles regarding relations and quantifiers that must be added to the classical probability calculus to obtain a reasonable calculus of nomic probabilities.

The calculus of nomic probabilities is based on the idea that nomic probability measures proportions among physically possible objects. The statistical law "$\mathbf{prob}(Fx/Gx) = r$" can be regarded as telling us that the proportion of physically possible *G*'s that would be *F* is *r*. Treating probabilities in terms of proportions is the key to understanding the logical and mathematical structure of nomic probability. It generates a number of useful theorems that are not theorems of the standard probability calculus. Here is one more example (reminiscent of Lewis' "principal principle"):

(PPROB) If *r* is a real number and $\Diamond[(\exists x)Gx\ \&\ \mathbf{prob}(Fx/Gx) = r]$ then $\mathbf{prob}\big(Fx\ /Gx\ \&\ \mathbf{prob}(Fx/Gx) = r\big) = r$.

Although the computational principles governing nomic probabilities are more complex than those embraced by the standard probability calculus, there is nothing philosophically suspect about them. They just constitute a slightly more elaborate calculus of probabilities that takes account of the fact that we are talking about indefinite probabilities rather than definite probabilities.

### 3.2 *The Statistical Syllogism*

Rules telling us when it is rational to believe something on the basis of high probability are called *acceptance rules*. The philosophical literature contains numerous proposals for acceptance rules, but most proceed in terms of definite probabilities rather than indefinite probabilities. There is, however, an obvious candidate for an acceptance rule that proceeds in terms of nomic probability. This is the *Statistical Syllogism*, whose traditional formulation is something like the following:

> **prob**$(Bx/Ax)$ is high.
> This is an $A$.
> ―――――――――――
> Therefore, this is a $B$.

It seems clear that we often reason in roughly this way. For instance, on what basis do I believe what I read in the newspaper? Certainly not that everything printed in the newspaper is true. No one believes that. But I do believe that *it is probable* that what is published in the newspaper is true, and that justifies me in believing individual newspaper reports. Similarly, I do not believe that every time a piece of chalk is dropped, it falls to the ground. Various circumstances can prevent that. It might be snatched in midair by a wayward hawk, or suspended in air by Brownian movement, or hoisted aloft by a sudden wind. None of these are at all likely, but they are possible. Consequently, all I can be confident of is that it is probable that chalk, when dropped, will fall to the ground. Nevertheless, when I drop a piece of chalk, I expect it to fall to the ground.

Clearly, the conclusion of the statistical syllogism does not follow deductively from the premises. The premises of the statistical syllogism can at most create a presumption in favor of the conclusion, and that presumption can be defeated by contrary information. In other words, the inference licensed by this rule must be a defeasible inference. The inference is a reasonable one in the absence of conflicting information, but it is possible to have conflicting information in the face of which the inference becomes unreasonable.

It turns out that to make the statistical syllogism work properly, it requires a qualification to the effect that the properties to which it appeals are projectible with respect to each other, in the familiar sense required by principles of induction. This is discussed further in the appendix.[8] The final version of the principle of the statistical induction that is assumed by the theory of nomic probability is the following:

(SS) If $Fx$ is projectible with respect to $Gx$ and $r > .5$, then "**prob**$(Fx/Gx) \geq r$" is a defeasible reason for the conditional "$Gc \rightarrow Fc$", the strength of the reason depending upon the value of $r$.

As with any defeasible inference, the use of (SS) can be defeated by having a reason for denying the conclusion. A reason for denying the conclusion constitutes a "rebutting defeater". But there is also an important kind of "undercutting defeater"—a defeater that attacks the inference itself rather than its conclusion.[9] In (SS), we infer the truth of "$Fc$" on the basis of probabilities conditional on a limited set of facts about $c$ (the facts expressed by "$Gc$"). But if we know additional facts about $c$ that lower the probability,

---

[8] For a useful compilation of articles on projectibility, see Stalker (1994).

[9] For a general account of defeasible reasoning, see my (1995).

that defeats the defeasible inference:

(D) If *Fx* is projectible with respect to (*Gx*&*Hx*) then "*Hc* & **prob**(*Fx*/*Gx*&*Hx*) < **prob**(*Fx*/*Gx*)" is an undercutting defeater for (SS).

The defeaters described by principle (D) are *subproperty defeaters.* (D) amounts to a kind of "total evidence requirement". It requires us to make our inference on the basis of the most comprehensive facts regarding which we know the requisite probabilities. (D) is not the only defeater required for (SS), but I will not discuss further defeaters here.[10]

I take it that (SS) is actually quite an intuitive acceptance rule. It amounts to a rule saying that, when *F* is projectible with respect to *G*, if we know that most *G*'s are *F*, that gives us a reason for thinking of any particular object that it is an *F* if it is a *G*. The only surprising feature of this rule is the projectibility constraint, but examples are given in the appendix illustrating why that constraint is necessary. (SS) is the basic epistemic principle from which all the rest of the theory of nomic probability is derived.

### 3.3 *Statistical Induction*

The values of some nomic probabilities are derivable from the values of others using the calculus of nomic probabilities, but our initial knowledge of nomic probabilities must result from empirical observation of the world around us. This is accomplished by *statistical induction.* We observe the relative frequency of *F*'s in a sample of *G*'s, and then infer that **prob**(*Fx* / *Gx*) is approximately equal to that relative frequency. One of the main accomplishments of the theory of nomic probability is the demonstration that precise principles of induction can be derived from (and hence justified on the basis of) the acceptance rule (SS) and computational principles constituting the calculus of nomic probabilities. In particular, the projectibility constraint in (SS) turns out to be the source of the familiar projectibility constraint on induction. This leads to a solution of sorts to the problem of induction. I will not go into any more detail here, but the derivation is sketched in the appendix.

### 3.4 *Non-Classical Direct Inference*

"Indifference principles" have played an important role in the history of probability theory. They tell us, roughly, that we can assume defeasibly that differences do not make a difference unless we have some reason for thinking otherwise. Laplace's theory was based heavily on such a principle (also called "the principle of insufficient reason"). However, his use of the principle is well known to be logically inconsistent.[11] Still, there is something intuitively tempting about such principles.

One of the most interesting features of the theory of nomic probability is that it entails a kind of indifference principle. This is not a primitive posit

---

[10] See my (1990).

[11] This is based upon what is known as "Barnard's paradox". See the discussion in chapter 1 of my (1990).

about nomic probabilities, but rather a theorem resulting from (SS) and the calculus of nomic probabilities. The principle consists of a defeasible principle of inference, and some defeaters for it. The *principle of non-classical direct inference* is the following:

(DI) If $F$ is projectible with respect to $G$ then "$\mathbf{prob}(Fx/Gx) = r$" is a defeasible reason for "$\mathbf{prob}(Fx/Gx\&Hx) = r$".

This principle is very important for the practical use of probabilities. When we employ probabilities in our reasoning, we assume that probabilities whose values we do not know would not upset that reasoning if we did know them. And that amounts to the assumption that adding conjuncts to the condition of a nomic probability will not usually change the value of the probability. It is gratifying then that (DI), which legitimizes this assumption, can be derived mathematically from the primitive assumptions of the theory of nomic probability.

Just as important as the principle (DI) is an array of defeaters for it. The simplest defeater is a variety of subproperty defeat analogous to subproperty defeat for (SS):

(SDI) If $F$ is projectible with respect to $J$ and $\Box(\forall x)[(Gx\&Hx) \rightarrow Jx]$ then "$\mathbf{prob}(Fx/Gx) \neq \mathbf{prob}(Fx/Gx\&Jx)$" is an undercutting defeater for (DI).

The intuitive idea here is that if $(Gx\&Hx)$ entails $Jx$, then $(Gx\&Hx)$ entails $(Gx\&Hx)$, which entails $Gx$. In this sense, logically, $(Gx\&Jx)$ "lies between" $Gx$ and $(Gx\&Hx)$. The presumption that $\mathbf{prob}(Fx/Gx) = \mathbf{prob}(Fx/Gx\&Hx)$ depends on the probability of $\mathbf{prob}(Fx/Gx)$ being propogated along that logical path. When it is not, that defeats the inference. As in the case of (SS), other defeaters are needed as well but I will not pursue that here.

## 4. Mixed Physical/Epistemic Probabilities

Our objective in this chapter is to find a kind of probability that is of use in calculating the expected values required by decision-theoretic reasoning. Nomic probabilities, being indefinite probabilities, cannot play that role. The requisite probabilities must be definite probabilities, telling us how likely specific outcomes are when we perform an action.

The probabilities used in practical deliberation must have a strong epistemic element. For example, if I am deciding whether to carry an umbrella when I go to work today, before looking outside I may know that the indefinite probability of rain in Tucson this time of year is only .05, and so I will conclude that it is extremely unlikely to rain today. But then I look out the window and see huge dark clouds looming overhead. At that point I conclude that rain is likely. The probability that I employ in deciding whether to carry an umbrella changes as my knowledge of the situation changes.

There are two possible construals of the way our knowledge of our circumstances affects our assessment of the definite probability. It could be that the probability itself changes when I acquire additional relevant knowledge, or it could be that the probability is fixed but my belief about its value changes. However, it can be argued that the second construal cannot be right. This turns on the observation that when I become completely justified in believing *P*, I assign it the value 1 for decision-theoretic purposes and I assign ~*P* the value 0. For example, consider a lottery that has already taken place. Ticket 345 was the winning ticket in the lottery and I know it. If I am offered the opportunity to purchase ticket 345, I may do so. However, I will assign other tickets the probability 0 of being the winning ticket, and I will not even consider purchasing one of them, no matter how cheap it is and no matter how valuable the lottery. Similarly, if I look out the window and see that it is presently raining, I will take the probability of rain today to be 1. If this is regarded as a discovery about what the value of the probability was all along, then we must conclude that all truths have probability 1 and all falsehoods have probability 0. In other words, it is a necessary truth that for any proposition *P*, either **PROB**(*P*) = 1 or **PROB**(*P*) = 0. If that is a necessary truth, then if *P* is some proposition whose truth value I do not know, I cannot reasonably conclude that it has a probability between 0 and 1. If the definite probabilities of use in decision-theoretic reasoning are to make sense, they cannot just reflect truth values. Accordingly, we must reject the suggestion that whenever I acquire additional relevant knowledge, my belief about the definite probability changes but the definite probability itself remains unchanged. We must instead conclude that the definite probability itself changes when I acquire additional relevant knowledge. The probability is a function in part of my epistemic state.

Clearly, the definite probability whose value we want to know is also sensitive to general meteorological probabilities relating clouds to rain. These are nomic probabilities. If the nomic meteorological probabilities were different, the likelihood of rain today would be different. We can express this by saying that the definite probabilities are *mixed physical/epistemic probabilities* —they are sensitive to both nomic probabilities and our epistemic situation. Henceforth, I will take **PROB**(*P*) to be the mixed physical/epistemic probability of the proposition *P*.

Just like subjective probabilities, mixed physical/epistemic probabilities must be relativized to cognizers. Because different cognizers have different information at their disposal, the mixed physical/epistemic probabilities of outcomes may be different for them. If my wife has yet to look out the window, for her the probability of rain today may be .05, while for me it is .75. However, unlike subjective probabilities, mixed physical/epistemic probabilities are objectively determined by the relevant nomic probabilities and the cognizer's knowledge of his circumstances.

What is characteristic of mixed physical/epistemic probabilities is the way we reason about them. We infer them by direct inference from nomic probabilities and information about the epistemic state of the cognizer. Let us look more closely at how direct inference works. This will lead us to an understanding of what these probabilities come to.

### *4.1 Classical Direct Inference*

The basic idea behind classical direct inference was first articulated by Hans Reichenbach (1949)—in determining the probability that an individual $c$ has a property $F$, we find the narrowest reference class $X$ to which $c$ belongs and for which we have reliable statistics and then infer that **PROB**$(Fc)$ = **prob**$(Fx/x\in X)$. To illustrate, suppose we want to know the probability that Anne, who is of European origin, is lactose intolerant. If we know that the probability of an adult of European origin being lactose intolerant is 0.2, and do not know any other relevant probabilities, then we will take the probability of Anne being lactose intolerant is 0.2. But suppose we also know that Anne is of Scandinavian origin, where the probability of an adult of Scandinavian origin being lactose intolerant is only 0.03. Then we will conclude instead that the probability of Anne being lactose intolerant is only 0.03.

There is almost universal agreement that direct inference is based upon some such principle as the one Reichenbach described, but there is little agreement about the precise form the theory should take. In my (1983) and (1984), I argued that classical direct inference should be regarded as proceeding in accordance with two epistemic principles. Where **K**$P$ means "$P$ is known (by the cognizer) to be true":

(CDI) If $F$ is projectible with respect to $G$, then "**prob**$(Fx/Gx) = r$ & $Gc$ & **K**$(P \leftrightarrow Fc)$" is a defeasible reason for "**PROB**$(P) = r$".

(SCDI) If $F$ is projectible with respect to $H$ then "**K**$Hc$ & **prob**$(Fx/Gx\&Hx) \neq$ **prob**$(Fx/Gx)$" is an undercutting defeater for (CDI).

Principle (SCDI) formulates a kind of subproperty defeat for direct inference, because it says that probabilities based upon more specific information take precedence over those based upon less specific information. The projectibility constraint in these rules is required to avoid various paradoxes of direct inference, but I will not discuss that here. See my (1990) for a full discussion of the matter.[12] Again, more defeaters are required to get a complete theory of direct inference.

To illustrate this account of direct inference, consider Anne again. If we know nothing relevant about Anne except that she is an adult of European descent, we will infer that the probability of her being lactose intolerant is 0.2. (CDI) provides us with a defeasible reason for making this inference. But if we also know that Anne is of Scandinavian origin, (CDI) provides us with a second defeasible reason, this time for the conclusion that the probability of Anne being lactose intolerant is 0.03. These two conclusions conflict, and intuitively we want to accept the second conclusion. This is justified by (SCDI), which provides a defeater for the inference to the conclusion that the probability is 0.2, leaving the other inference undefeated.

[12] For competing theories, see Kyburg (1974), Levi (1977, 1980, 1981), Bacchus (1990), Halpern (1990).

### *4.2 Reducing Classical Direct Inference to Non-Classical Direct Inference*

We now have two kinds of direct inference — classical and non-classical. Direct inference has traditionally been identified with classical direct inference, but I believe that it is most fundamentally non-classical direct inference. The details of classical direct inference are all reflected in non-classical direct inference. If we could identify definite probabilities with certain indefinite probabilities, we could derive the theory of classical direct inference from the theory of non-classical direct inference. This can be done as follows. Let $\mathbf{K}$ be the conjunction of all the propositions known by the cognizer to be true. For any proposition $P$, let *P-worlds* be physically possible worlds at which $P$ is true. Then using the same proportion function that was employed in the definition of nomic probabilities, we can define

> $\mathrm{PROB}(P)$ is the proportion of $\mathbf{K}$-worlds that are also $P$-worlds.
>
> If $Q$ is consistent with $\mathbf{K}$ then $\mathrm{PROB}(P/Q)$ is the proportion of $(Q\&\mathbf{K})$-worlds that are also $P$-worlds.

This is a natural way of understanding definite probabilities, and the theory of nomic probabilities then enables us to prove the following representation theorem:

$$\mathrm{PROB}(Fa) = \mathbf{prob}(Fx \;/\; x = a \;\&\; \mathbf{K}).$$

This is the probability of something having the property $F$ given that it is the object $a$ and given everything we know. Technically, this is an indefinite probability, but it is only about the object $a$. Presumably, this nomic probability can only be evaluated by using non-classical direct inference, and the structure of that reasoning mirrors the structure of the reasoning involved intuitively in classical direct inference. More generally, we can prove:

> If it is physically necessary that $[\mathbf{K} \rightarrow (Q \leftrightarrow Sa_1\ldots a_n)]$ and that $[(Q\&\mathbf{K}) \rightarrow (P \leftrightarrow Ra_1\ldots a_n)]$, and $Q$ is consistent with $\mathbf{K}$, then $\mathrm{PROB}(P/Q) = \mathbf{prob}(Rx_1\ldots x_n \;/\; Sx_1\ldots x_n \;\&\; x_1 = a_1 \;\&\; \ldots \;\&\; x_n = a_n \;\&\; \mathbf{K})$.

This makes precise the way in which the definite probabilities at which we arrive by direct inference are mixed physical/epistemic probabilities. We obtain definite probabilities by considering indefinite probabilities conditional on properties we are warranted in believing the objects in question to possess.

### *4.3 Betting the Farm*

On the preceding account, if you are warranted in believing $P$, then $\mathrm{PROB}(P) = 1$. But then it follows that you can simply assume the truth of $P$ in decision-theoretic reasoning. In particular, if you are offered a bet which returns one cent if $P$ is true and costs \$$x$ if $P$ is false, that bet has a positive expected value regardless of what $x$ is, and so you should accept it. But to many philosophers, this seems dubious. There is a temptation to say that no matter how warranted we regard ourselves as being in believing $P$, we would not bet the farm on $P$. Does that show anything about PROB($P$)?

Imagine that I show you a quarter. You look at it and it, handle it, and so on, and it seems plain to you that it is a quarter. If I ask you (in an uncontentious manner) what the probability is of its being a quarter, you are apt to respond, "What do you mean 'probability'? It's a quarter. It's certain, not probable." But then I offer to bet that it is not really a quarter. I offer a bet of \$100—if it is a quarter, you win \$100, and if it is not then I win \$100. Would you bet? You might not. The very fact that I offer you the bet makes you suspicious that there is some trick involved. However, this is a good bet as long as the probability of its being a quarter is greater than .5. Surely, prior to my offering you the bet, you would regard the probability that it is a quarter as much greater than .5. So what bets of this sort you would accept does not accurately measure antecedent probability judgments.

When the act of offering the bet does not make you suspicious, would you bet the farm on it? Of course you would. Suppose it really is the farm that is at stake. I hold the mortgage on your farm. In a gesture of good will, I offer to sell it back to you for a quarter. Unless you suspect that something fishy is going on, you would not hesitate to give me that quarter in exchange for the farm.

But now make the case slightly more difficult. My offer really is on the up and up. I am a good guy and you know it. But I also have a quick temper. I made a similar altruistic offer to someone else a year ago, and he had the temerity to try to pay me with a fake quarter (a rather bad fake at that). When I realized what he was doing, my temper exploded and rather than selling him the farm I foreclosed on his mortgage and he lost his farm.

Under the circumstances, you want to be sure that the coin you give me is not a fake quarter. But if you examine it carefully, you will be confident of that, and you will give me the quarter in exchange for the farm. In doing this you are, in effect, placing a bet. Consider the logical form of this bet and what we can conclude from the fact that you will accept it and are seemingly rational in doing so. You believe ($P$) that the coin is a quarter, and in fact assign a probability of 1 to it. Now you must decide whether to give me the coin in response to my offer. If you give me the coin and it is a quarter, you get the farm. If you give me the coin and it is not a quarter, you lose the farm. If you do not give me the quarter, you get the status quo, which is that you must continue paying the mortgage, but will eventually pay if off and get the title to the farm. So the expected value of giving me the coin is PROB($P$)$\cdot$*farm* – 25¢. The expected value of not giving me the coin is *farm* – *payoff*. So giving me the coin is reasonable iff

$$\text{PROB}(P) > 1 - (\textit{payoff} - 25¢)/\textit{farm}.$$

I have argued that no matter how much we make the value of the farm and how little we make the payoff (greater than 25¢), after carefully examining the quarter, you would give it to me rather than continuing to make the mortgage payments. But for that to be rational, PROB($P$) must be 1.

We bet our *lives* on justified beliefs every day. When I take a bite of food I am betting that it has not been poisoned and that it is not infected with botulism. If we suppose the alternative to taking the bite of food is to be mildly hungry, then the logical form of the decision is the same.

For another example, think about the preflight check on an airliner. The pilot checks many things. Among them is the fuel level. If the plane does not have sufficient fuel, it will crash. But the pilot accepts the result of his check. He is betting his life and the life of his passengers on this. Is this irrational? Surely not. Do you think that rationality dictates never flying on airplanes?

What about the example, discussed in chapter 6, of the car keys? I believe that my car keys are on my dresser. But I would probably not bet the farm on that. Why not? I have mislaid my car keys too often. In this example, it seems that I initially do believe that my keys are on my dresser, but when the issue of betting arises I reconsider my belief and decide that my initial belief was not really justified. All I should have believed was that it is fairly probable that my keys are on the dresser. So I decline to bet. This case is thus different from the cases discussed above.

The thought that justified beliefs should not have probability 1 because we would not bet the farm on them is just wrong. We bet the farm, and in fact, our lives, on justified beliefs every day. That is what justified beliefs are for. That give us the background of information that we can simply assume in the course of rational deliberation.

## 5. Conclusions

Once we give up the old-fashioned idea that objective probability makes no sense unless it can be defined in terms of relative frequencies, the concept becomes no more philosophically suspect than other philosophically interesting concepts. It turns out that it can be given an epistemological analysis. That is what the theory of nomic probability accomplishes. Beginning with the statistical syllogism and a fairly minimal set of assumptions about the calculus of nomic probability, we can derive principles of inductive reasoning and principles of direct inference. In particular, we can define mixed physical/epistemic probabilities in terms of nomic probabilities.

The question remains whether these are the appropriate probabilities for use in computing expected values. I used to think they are. They at least combine many of the features we want such probabilities to have. But there is a residual difficulty. This has to do with Newcomb's problem and the related problems that gave rise to "causal decision theory". This will be taken up in the next chapter, where it is shown how to define a kind of "causal probability" in terms of mixed physical/epistemic probabilities. That, finally, will be the probability that I believe to be of use in decision-theoretic reasoning.

# 8

# Causal Probabilities

## 1. Causal Decision Theory

It is natural to suppose that the mixed physical/epistemic probabilities defined in chapter 7 are the appropriate probabilities for use in defining expected values, as used in the optimality prescription. The result would be to define expected values as follows:

$$\mathbf{EV}(A) = \sum_{O \in \mathbf{O}} \mathrm{U}(O) \cdot \mathbf{PROB}(O/A).$$

However, Nozick (1969) presented what has come to be known as "Newcomb's problem", and that has led to a general recognition that when decision theory is based on such a probability it can make incorrect prescriptions in some cases. The Newcomb problem itself commands conflicting intuitions, so I will not discuss it here.[1] There are other examples that are clearer. One of the more compelling examples is due to Stalnaker (1978). Suppose you are deciding whether to smoke. Suppose you know that smoking is somewhat pleasurable, and harmless. However, there is also a "smoking gene" present in many people, and that gene both (1) causes them to desire to smoke and (2) predisposes them to get cancer (but not by smoking). Smoking is evidence that one has the smoking gene, and so it raises the probability that one will get cancer. Getting cancer more than outweighs the pleasure one will get from smoking, so when expected values are defined as above, the optimality prescription recommends not smoking. But this seems clearly wrong. Smoking does not *cause* cancer. It is just evidence that one already has the smoking gene and hence may get cancer from that. If you have the smoking gene, you will still have it even if you refrain from smoking, so the latter will not prevent your getting cancer. As Joyce (1998, p. 146) remarks, "Rational agents choose acts on the basis of their *causal efficacy*, not their auspiciousness; they act to *bring about* good results even when doing so might betoken bad news."

As a number of authors (Gibbard and Harper 1978; Sobel 1978, 1994; Skyrms 1980, 1982, 1984; Lewis 1981) have observed, conditional probabilities can reflect either evidential connections or causal connections. In this example, the connection between smoking and getting cancer is merely evidential. Smoking is evidence for cancer, but it does not cause it. In deciding whether

---

[1] For an excellent discussion of the Newcomb problem, see Joyce (1998), pp. 146 ff.

to perform an action, we consider the consequences of performing it. The consequences should be its *causal consequences*, not its evidential consequences. This suggests that a correct formulation of decision theory should replace the conditional probability **PROB**($O/A$) by some kind of "causal probability". The resulting theories are called *causal decision theories* and are contrasted with *evidential decision theory*.

Several authors have tried to argue that evidential decision theory can be understood in such a way that it does not yield the counter-intuitive conclusions it seems to yield in examples like the above, in effect arguing that it yields the same conclusions as causal decision theories. But I will postpone discussing these attempts until section 6. For now I will assume that evidential decision theory works as just described.

A natural first attempt at formulating causal decision theory would appeal to the literature on probabilistic causation. I will begin by discussing that, but I will dismiss it as inapplicable. Then I will discuss Skyrms' proposal, and Lewis' emendation to it. That leads to a causal probability **K-PROB**. Skyrms shows that by appealing to **K-PROB** we can avoid the smoking gene problem. However, **K-PROB** is defined in terms of an undefined concept of causal independence. I propose to avoid that by defining a different causal probability **C-PROB**. **C-PROB** is defined in terms of temporal relations rather than causal independence, and is shown to handle the smoking gene problem in the same way. It is argued that if two plausible assumptions are made, **C-PROB** and **K-PROB** are equivalent. Then I will investigate the mathematical properties of **C-PROB** and show that there are efficient ways of computing its values. In fact, it will often be easier to compute the value of **C-PROB** than it is to compute the value of the mixed physical/epistemic probability **PROB**.

## 2. Probabilistic Causation

There is a substantial literature on probabilistic causation, and it would be natural to suppose that causal decision theory should be formulated in terms of the kind of probabilistic causation discussed there. It should be acknowledge from the outset that this work has not generally been directed at decision theory. The aim of the work is different. As far as I know, the only one who one has seriously suggested using it for computing expected values is Pearl (2000).[2] And as I will now explain, I believe that there are good reasons for thinking that probabilistic causation cannot serve the purposes of causal decision theory.

The literature on probabilistic causation has two strands. On the one

---

[2] Joyce (1998) makes remarks that *sound* sympathetic to this view. He asserts (p. 161) that causal probabilities are the agent's estimates of "causal tendencies". This sounds like an indefinite causal probability (see below). However, when he works out the details of his theory, this is not the direction he goes. His approach is closer to that of Gibbard and Harper (1978).

hand, it is concerned with the analysis of claims like "Smoking causes cancer", where this does not mean that smoking *always* causes cancer. On the other hand, it is concerned with how one can use statistical data (correlations) to discover causal connections. The targeted causal connections are the probabilistic causes that are the focus of the first strand.

Nancy Cartwright (1979) made the observation years ago that probabilistic causation is most naturally understood in terms of the indefinite probability of deterministic causation. For example, to say that smoking causes cancer is to say that the probability of a person's smoking causing him to get cancer is above some (contextually determined) threshold. The relevant probability is the indefinite probability **prob**($x$'s smoking causes $x$ to get cancer/ $x$ smokes). This accommodates the fact that in some people smoking causes cancer, and in others it does not. Cartwright's account has always seemed to me to be right.

If we understand probabilistic causation in this way, it is not directly of use in causal decision theory, because for decision making we need a definite probability. However, by direct inference from such indefinite probabilities we can infer definite probabilities of the form PROB($A$ causes $O/A$), where $A$ is an action and $O$ is a possible outcome. Is this the probability we should be using in decision making? It seems to me that it is not. The difficulty is that causation has too rich a logical structure for our purposes. For example, it is generally agreed that in cases of causal overdetermination, neither of the "overdeterminers" causes the effect. For example, if Jones and Smith shoot Robinson simultaneously and Robinson dies as a result of it, it is not true that Jones' shooting Robinson caused Robinson to die. This is because Robinson would have died anyway (because Smith shot him). Although this seems to be a correct observation about causes, it does not seem to be relevant to the problems that give rise to causal decision theory. For instance, imagine that smoking has the potential to cause cancer, but so does heavy drinking. Furthermore, smokers tend to be heavy drinkers. Suppose that in most cases in which smokers get cancer this is overdetermined by their both smoking and drinking. Thus, although smoking is in some sense "causally involved" in getting cancer, in most cases the probability of smoking causing cancer is low. This, however, should not have the consequence that it is all right to smoke. This difficulty is just a reflection of the logical complexity of the concept of a cause. For the purposes of decision making, what is relevant is a weaker notion of "causal connection".

One possibility is that expected values should be computed in terms of probabilities of the form PROB($A$ causes $O$ or $A$ overdetermines $O/A$). However, this seems to let in too much. Suppose that smoking can cause cancer. The causal mechanism by which it does this involves causing genetic damage that inserts a certain sequence into the smoker's DNA. Having that sequence in one's DNA both disposes one to get lung cancer and disposes one to smoke if one doesn't already smoke. Some people are born with this sequence in their DNA, and others acquire it as a result of smoking, which is a potent cause of it. It seems then that smoking either causes or overdetermines having that sequence in one's DNA. In particular, if one was born with the sequence, smoking has a tendency to overdetermine having the sequence. We can suppose that having that sequence makes it reasonably likely that one will get cancer. Should one smoke if one desires to? That depends on the probability that if you desire to smoke then you were born with the sequence in your DNA. Suppose that is extremely likely. Then if you get a lot of pleasure out of smoking, you might as well smoke. But if it is rare for people to be born with the sequence in their DNA, then you should not smoke. However, in either case smoking tends to either cause or causally overdetermine subsequently having cancer, so we cannot distinguish between these two cases by appealing to that weaker kind of causal connection.

Apparently the relevant probability to use in decision making is neither the definite probability of the action causing an outcome nor the definite probability of the action being more weakly causally connected to it. We need a different kind of probability. Can the literature on probabilistic causation help us out by providing a different kind of probabilistic causal connection? Most of the work on probabilistic causation stems originally from Suppes (1970, 1988). More recent work can be found in Glymour (1988), Glymour and Cooper (1998), Meek and Glymour (1994), Pearl (2000), and Spirites, Glymour, and Scheines (1993). This work has the double aim of analyzing probabilistic causation without appealing to causation simpliciter (unlike Cartwright's proposal) and of investigating how one can use statistical information to detect causes.

It is striking that all of this work on probabilistic causation begins by giving examples analogous to "Smoking causes cancer". For instance, Suppes (1970) gives the example "Reckless driving causes accidents", and Pearl (2000) mentions this example with approval. These are not "single case" probability statements. They are most naturally understood as indefinite probability statements about causes. But it is equally striking that, with the exception of Cartwright (1979), none of these authors employ indefinite probabilities in their analyses. Suppes (1973) appeals to "single-case propensities", which are supposed to be objective definite probabilities. Glymour and Pearl do not says what kind of probability they have in mind, but it is clearly some kind of definite probability.[3] This creates a problem for all of these theories, because there does not seem to be any suitable kind of definite probability for use by these theories in analyzing causation that is also of use in decision making. This turns on the distinction between objective and (at least) partly subjective probabilities. In the latter category I have talked about subjective probability and mixed physical/epistemic probability. In the literature on probability theory, there is also a discussion of various kinds of completely objective definite probabilities. These are sometimes called "objective chances" or "propensities" (see for example Fetzer 1971, 1977, 1981; Giere 1973, 1973a, 1976; Mellor 1969, 1971; Suppes 1973). I proposed in my (1990) to analyze the objective chance of an event as the nomic proba-

---

[3] When I asked Glymour about this, he said that the official view of him and his co-authors was that "we don't care, as long as the joint probability distribution satisfies certain assumptions."

bility of such an event occurring conditional on everything true in the past light cone of the event.

Eells (1991) observes that if our objective is to *analyze* causation in terms of probabilities, it is difficult to see why one would suppose that this can be done in terms of probabilities that are not fully objective. Whether *P* causes *Q* is a purely objective matter of fact, independent of what anyone knows or believes, so how could it possibly be analyzable in terms of either a purely subjective or a partly subjective variety of probability?

On the other hand, if one analyzes probabilistic causation in terms of purely objective probabilities, two other difficulties arise. First, it is generally agreed in the literature on objective definite probabilities that they only have values intermediate between 0 and 1 in non-deterministic worlds. In a deterministic world, the objective probability of a proposition is 1 iff it is true (Giere 1973). On all analyses of probabilistic causation, causes must raise the probabilities of their effects. But in a deterministic world nothing could raise the probability of a true proposition so it again follows that there are no probabilistic causes in deterministic worlds. However, for rational decision making, we need a kind of probability that enables us to define expected values even in deterministic worlds.

There is a related difficulty for the use of probabilistic causation in decision theory. If probabilistic causation is analyzed in terms of purely objective probabilities, then the resulting causal probabilities do not reflect the agent's knowledge or beliefs. That is just what it is for them to be objective. However, rational decisions must always be made by reference to the agent's knowledge or beliefs. We saw in chapter 7 (section 4) that this requires the probabilities used in computing expected values to be partly subjective. It seems to follow that even if a correct analysis of probabilistic causation is forthcoming from an appeal to objective probabilities, it cannot play the desired role in causal decision theory.[4]

To recapitulate, existing theories of probabilistic causation have no hope of being right unless they appeal to purely objective definite probabilities. But if they do, they cannot produce causal probabilities that take account of decision makers' knowledge. Why has this been overlooked? I think that much of this literature suffers from the common conflation of definite and indefinite probabilities. For some reason, very few philosophers or contemporary probability theorists take any note of indefinite probabilities. I suspect that much of the work on using statistical information to detect causal connections can be given a firm foundation by reformulating it in terms of indefinite probabilities. In that case it becomes applicable to deterministic and indeterministic cases alike. But even if that is right, the resulting probabilities are indefinite, not definite. Hence they cannot be the probabilities we need for causal decision theory.

My conclusion is that we must look beyond the work on probabilistic causality for a variety of probability of use in causal decision theory. I turn

---

[4] Hitchcock (1996) makes this same observation.

next to the theories of Brian Skyrms and David Lewis.

## 3. Skyrms and Lewis

Most of the work on causal decision theory has been carried out within the framework of subjective probabilities. However, in some cases the basic ideas are largely independent of that, and can be reformulated in terms of non-subjective probabilities.[5] In this connection, let us consider Brian Skyrms' (1980, 1982, 1984) proposal. Skyrms suggests distinguishing between the *background* of an action (my terminology) and the *consequences* of an action. The background consists of situations (types of world-states) that are causally independent of the performance of the action. That is, the action does not cause or causally dispose them to occur or not occur. Let us call these *K-backgrounds*. The consequences of an action can then be evaluated against a background $K$ by considering the probability $\mathbf{PROB}(O/A\&K)$. In computing the expected value of an action, some parts of the background will consist of things we know to be true, but other parts of the background may be unknown to us, having only probabilities associated with them. Skyrms' suggestion is that if there is a finite set $\mathbb{K}$ of backgrounds that we consider possible, then in computing the expected value we consider the probability of an outcome relative to each possible background, and weight it by the probability of that background being true. In other words, we can define causal probability as follows:

$$\mathbf{K\text{-}PROB}_A(O) = \sum_{K\in\mathbb{K}} \mathbf{PROB}(K)\cdot\mathbf{PROB}(O/A\&K).$$

This makes causal probability the mathematical expectation of the probability of the outcome on the different possible backgrounds. It is easily verified that $\mathbf{K\text{-}PROB}_A$ is a probability, that is, it satisfies the probability calculus. Then the proposal is to define expected value in terms of $\mathbf{K\text{-}PROB}_A(O)$ instead of $\mathbf{PROB}(O/A)$:

$$\mathbf{EV}(A) = \sum_{O\in\mathbb{O}} \mathrm{U}(O)\cdot\mathbf{K\text{-}PROB}_A(O).$$

The $K$-backgrounds constitute a partition. That is, they are mutually exclusive and their disjunction is a necessary truth. Skyrms describes the $K$-backgrounds as maximally specific specifications of factors outside the agent's influence (at the time of the decision) that are causally relevant to the outcome of the

---

[5] The approach of Gibbard and Harper (1978) identifies causal probability with the subjective probability of a "causal" subjunctive conditional, and it can probably not be reformulated in terms of non-subjective probabilities. This idea is generalyzed by Joyce (1998), who discusses what are in effect subjunctive subjective probabilities. In light of the conclusions of chapter 6 regarding the untintelligibility of subjective probability, I will not discuss that approach further.

agent's action.

Let us see how this proposal handles the smoking gene example. Whether the person has the gene or not ($G$ or $\sim G$) is outside his influence. If he has it, he already has it when he makes his decision whether to smoke, so that decision cannot causally influence his having or not having the gene. If $G$ and $\sim G$ are the only elements of the background causally relevant to his getting cancer, and nothing unknown to him is relevant to his getting pleasure from smoking, then the expected value of smoking ($S$) can be computed as follows:

$$\begin{aligned}\mathbf{EV}(S) &= \mathbf{U}(\textit{pleasure of smoking})\cdot\textbf{K-PROB}_S(\textit{pleasure of smoking}) \\ &\quad + \mathbf{U}(\textit{cancer})\cdot\textbf{K-PROB}_S(\textit{cancer}) \\ &= \mathbf{U}(\textit{pleasure of smoking})\cdot\textbf{PROB}(\textit{pleasure of smoking} / S) \\ &\quad + \mathbf{U}(\textit{cancer})\cdot[\textbf{PROB}(G)\cdot\textbf{PROB}(\textit{cancer} / S\&G) \\ &\qquad + \textbf{PROB}(\sim G)\cdot\textbf{PROB}(\textit{cancer} / S\&\sim G)].\end{aligned}$$

We have made the assumption that $\textbf{PROB}(\textit{cancer}/S\&G) = \textbf{PROB}(\textit{cancer}/G)$ and $\textbf{PROB}(\textit{cancer}/S\&\sim G) = \textbf{PROB}(\textit{cancer}/\sim G)$, so:

$$\begin{aligned}\mathbf{EV}(S) &= \mathbf{U}(\textit{pleasure of smoking})\cdot\textbf{PROB}(\textit{pleasure of smoking}/S) \\ &\quad + \mathbf{U}(\textit{cancer})\cdot[\textbf{PROB}(G)\cdot\textbf{PROB}(\textit{cancer}/G) \\ &\qquad + \textbf{PROB}(\sim G)\cdot\textbf{PROB}(\textit{cancer}/\sim G)].\end{aligned}$$

Similarly, the expected value of not smoking ($\sim S$) is:

$$\mathbf{EV}(\sim S) = \mathbf{U}(\textit{cancer})\cdot[\textbf{PROB}(G)\cdot\textbf{PROB}(\textit{cancer}/G) + \textbf{PROB}(\sim G)\cdot\textbf{PROB}(\textit{cancer}/\sim G)].$$

Thus if $\mathbf{U}(\textit{pleasure of smoking}) > 0$ and $\textbf{PROB}(\textit{pleasure of smoking}/S) > 0$, it follows that $\mathbf{EV}(S) > \mathbf{EV}(\sim S)$. So causal decision theory recommends smoking, which is the right choice.

David Lewis (1981) endorses a causal decision theory with the same form as Skyrms', and represents that general form as the fundamental idea behind all causal decision theories. The difference between his theory and Skyrms' is that he takes the $K$'s to be what he calls *dependency hypotheses* — maximally specific propositions about how things the agent cares about do and do not depend causally on his present actions. Lewis proposes a narrow and a broad reading of Skyrms. On the narrow reading, the background $K$ consists of propositions describing singular states of affairs in the world (for example, the agent does or does not have the cancer gene). On the broad reading, backgrounds are the same as Lewis's dependency hypotheses. Lewis observes that on the broad reading, his theory is the same as Skyrms. He goes on to argue that what he regards as the other major causal decision theories (Gibbard and Harper 1978, Sobel 1978) are also equivalent to the

Skyrms/Lewis theory on at least some interpretation.[6]

It is not clear to me from Lewis' definition whether dependency hypotheses are supposed to include a specification of which of the singular states of affairs that are causally independent of the action are true. However, if we do not include that, there is no apparent way to justify the calculation that is supposed to solve the smoking gene problem. For that calculation to work, *G* and ~*G* must be contained in dependency hypotheses. So I will interpret Lewis that way.

A close reading of Skyrms suggests that he did not intend Lewis' broad reading of his theory. On the contrary, *K*-backgrounds were supposed to consist of singular states of affairs. Lewis raises two objections to Skyrms' theory on this narrow reading. The first is that the probability of getting outcome *O* by performing action *A* depends not just on singular states of affairs, but also on laws of nature. Thus these must be included in the background. This is indeed a problem for the subjectivist (that is, the official) version of Skyrms' theory and might reasonably be taken to motivate an expansion of the *K*-backgrounds to make them look more like Lewis' dependency hypothesis. However, this is not a problem when the theory is formulated in terms of mixed physical/epistemic probabilities. It is a theorem of the calculus of nomic probabilities that causal laws have probability 1. Thus there is no need for a non-subjective causal decision theory to include them in the background. Conditionalizing on something with probability 1 cannot change the probability.

At this point the subjectivist is bound to object, "But we may be uncertain whether something is a law, so we have to attach probabilities to that and factor them into the computation of the causal probability." However, epistemic uncertainty is only represented probabilistically if you are a subjectivist. And we observed in chapter 6 that even if we pretend that subjective probabilities make sense, it is logically problematic to take them to represent degrees of justification. Numerous arguments throughout the epistemological literature demonstrate that epistemic degrees of justification do not conform to the probability calculus.

Instead of trying to factor uncertainty about the relevant laws and nomic probabilities into the probabilities used in computing expected values, we should acknowledge that such uncertainty will make one uncertain about the computation of expected values. If we are sufficiently uncertain, for example, we cannot even locate the relevant nomic probabilities within useful intervals, then we will not be able to draw justified conclusions about the expected values of our actions, and so our decision problem is not an instance of decision making under risk. It is instead an instance of decision making under uncertainty, and falls outside the purview of the optimality

---

[6] This claim is dubious. The theory of Gibbard and Harper requires a "causal" subjunctive conditional that satisfies conditional excluded middle, i.e., $(P \,\Box\!\!\rightarrow Q) \vee (P \,\Box\!\!\rightarrow {\sim}Q)$. Most authors seem to agree that there is no such conditional (see the discussion in chapter 5 of Joyce (1998)), and Gibbard and Harper themselves acknowledge that in cases in which conditional excluded middle fails, some other treatment of causal probability is required.

prescription. For instance, suppose I show you a button, and tell you that if you push it good things might happen, but also bad things might happen, and I have no idea what the probability of either is. Can you make a rational choice between pushing or not pushing the button? Surely not. You might choose at random to either push the button or not push the button, but there is no basis for rationally preferring one choice to the other.

Lewis raises another objection to the narrow construal that is more telling. He observes that the *K*'s are characterized in terms of causal dependence, but cognizers may be uncertain about causal dependence and may only attach probabilities to hypotheses about causal dependence. He suggests that these probabilities should somehow enter into the computation of causal probability. Skyrms (1980) takes this objection seriously, and suggests a modification of his theory that is intended to accommodate it.

I think that this objection is indicative of a more serious objection that is telling against all current versions of causal decision theory, including Lewis's. These theories are all formulated in terms of causal dependence or some similar notion, but no analysis is proposed for what this amounts to. Harking back to some of the observations made in section 2, to say that *P* is causally dependent on *Q* might be taken to mean that *P* causes (is a cause of) *Q*, or it might mean that *P* is either a cause or a causal overdeterminer of *Q*, or it might mean that *P* is a single-case probabilistic cause of *Q* (whatever that amounts to), or it might mean that states of affairs "like" *P* sometimes (occasionally, often) cause states of affairs like *Q*, and so on. In my estimation, no theory that takes causal dependence as primitive can throw adequate light on the structure of the probability reasoning required for rational choice. Thus I turn in the next section to an alternative analysis of causal probability that is not subject to this objection.

## 4. Defining Causal Probability

My objective is to find a way of defining causal probability that does not appeal to concepts like causation or causal dependence. The basic idea behind my proposal is simple—causal probability propagates forward in time, never backward.[7] My suggestion is that in computing the possible effects of an action, we think of the world as evolving causally over time, interject the action into the world at the appropriate point, and then propagate changes forward in time as the world continues to evolve. This way of conceptualizing the world as evolving in temporal order is precisely the same idea that underlies most current solutions to the Frame Problem in AI (see Shoham 1986, 1987; Hanks and McDermott 1986, 1987; Lifschitz 1987; Gelfond and Lifschitz 1993; Shanahan 1990, 1995, 1996, 1997; Pollock 1998a). Those solutions are based upon the idea that given a set of deterministic causal laws, to compute the result of a sequence of actions we imagine them

---

[7] This is reminiscent of Lewis' (1979) "non-backtracking conditionals".

occurring in temporal order and propagate the changes through the world in that order. I will say more about this below. As I will define causal probability, it does the same thing probabilistically.

To make this precise, let us make the simplifying assumption that actions occur instantaneously. They have dates that are single instants of time. These are *point-dated* actions.[8] Singular states of affairs also have dates, but I will allow them to be either time intervals or time instants (degenerate intervals). I also assume that we can assign dates to logical combinations built out of conjunctions, disjunctions, and negations of singular states of affairs. The date of a negation is the date of what it negates, the date of a conjunction is the union of the dates of the conjuncts, and the date of a disjunction is the union of the dates of the disjuncts. The date of such a combination can be a time interval with gaps. I will refer to these logical combinations of singular states of affairs as *states of affairs* simpliciter (dropping "singular"). Let us say that *Q postdates P* iff every time in the date of *P* (an interval) is earlier than every time in the date of *Q*. Let us say that *P predates Q* iff every time in the date of *P* is earlier than or the same as every time in the date of *Q*. So if a state predates a point-dated action, the end-point of its date may be the same as the date of the action. But if it postdates the action, it occurs wholly after the date of the action.

Now let us begin by considering the case in which the world is deterministic. This means that each complete state of the world determines each subsequent state. The determination is by physical laws. Each state nomically implies subsequent states. In asking whether a possible outcome would result from a particular world state in which an action is performed, we are asking whether the outcome will be present in subsequent states. In a deterministic world, *O* will result just in case the actual state of the world up to and including the time *A* is performed includes a set *B* of singular states of affairs such that *A&B* nomically implies *O*. I will call *B* a *background state* for *O* relative to *A*.

If we are uncertain about the precise state of the world, then we may be uncertain about whether *O* will result. The *causal probability* that *O* will result should be identified with the probability that the state of the world at the time *A* is performed contains a background state for *O* relative to *A*. If *B* is the only background state for *O* relative to *A*, then the causal probability of *O* given *A* should be identified with **PROB**(*B*). If instead there are a finite number of background states $B_1,\ldots,B_n$, then the causal probability of *O* given *A* should be identified with **PROB**($B_1 \vee \ldots \vee B_n$). Let us write this probability as **C-PROB**$_A(O)$.

**C-PROB**$_A(O)$ need not be the same as **PROB**($O/A$). The latter would be **PROB**($B_1 \vee \ldots \vee B_n/A$). *A* cannot *cause* changes to the background state, but it can be evidence regarding whether a background state is actual. This is precisely what happens in the smoking gene example. If we suppose that the gene causes cancer deterministically, then *G* is the sole background

[8] In my (2002), I discussed ways of relaxing this assumption.

state and $\text{PROB}(G/A) \neq \text{PROB}(G)$. The probability $\text{C-PROB}_A(O)$ is then equal to $\text{PROB}(G)$ rather than $\text{PROB}(G/A)$. This is a causal probability that results from propagating the effects of actions forward in time but not backward in time. We hold the background state fixed, assigning to background states whatever probabilities they have prior to the action's being performed.

If we turn to nondeterministic worlds, the background states may no longer nomically imply the outcomes. They may only confer probabilities on the outcomes. If there were a single background $B$ such that $O$ can only result from $A$ by having $B$ true, we could define

$$\text{C-PROB}_A(O) = \text{PROB}(B) \cdot \text{PROB}(O/A\&B).$$

More generally, if we could confine our attention to a finite set $\mathbf{B}$ of (pairwise logically disjoint) backgrounds, we could define:

$$\text{C-PROB}_A(O) = \sum_{B \in \mathbf{B}} \text{PROB}(B) \cdot \text{PROB}(O/A\&B).$$

That is, the causal probability is the mathematical expectation of the probability of the outcome on the different possible backgrounds.

To define $\text{C-PROB}_A(O)$ generally (when $O$ postdates $A$ and $A$ is a point-dated action), let $\mathbf{C}$ be the set of all possible singular states of affairs and negations of singular states of affairs predating $A$. Define an *initial A-world-state* to be a maximal subset of $\mathbf{C}$ nomically consistent with $A$. This is a complete description of the world state up to the time $A$ is performed. I will not usually distinguish between an initial $A$-world-state and the conjunction of its members. Let $\mathbf{W}$ be the set of all initial $A$-world-states. Then we can define $\text{C-PROB}_A(O)$ to be the mathematical expectation of the probability of the outcome on the different possible initial $A$-world-states. If $\mathbf{W}$ is finite, our definition becomes:

$$\text{C-PROB}_A(O) = \sum_{W \in \mathbf{W}} \text{PROB}(W) \cdot \text{PROB}(O/A\&W).$$

Realistically, $\mathbf{W}$ will be infinite, in which case $\text{C-PROB}_A(O)$ must be defined using the more general integral definition of expected value. Using the notation introduced in chapter 1:

$$\text{C-PROB}_A(O) = \text{EXP}\big(\text{PROB}(O/A\&W) \leq r) \;/\; W \in \mathbf{W}\big)$$

However, to keep the mathematics simple, I will pretend that $\mathbf{W}$ is finite and use the summation version of the definition. This will make no difference to the results. Because the set $\mathbf{W}$ is chosen independently of $O$, it is trivial to verify that $\text{C-PROB}_A$ is a probability, that is, that it satisfies the probability calculus.

Thus far, $\text{C-PROB}_A(O)$ has been defined for all states of affairs postdating $A$. It will be convenient to define $\text{C-PROB}_A(O)$ for a broader class of states of affairs, including states of affairs that do not postdate $A$. If $O$ predates $A$ we can stipulate:

$$\text{C-PROB}_A(O) = \text{PROB}(O).$$

If $O_1$ postdates $A$ and $O_2$ predates $A$, then we will further stipulate that

$$\text{C-PROB}_A(O_1 \,\&\, O_2) = \text{PROB}(O_1)\cdot\text{C-PROB}_A(O_2/O_1).$$

However, this definition proceeds in terms of conditional causal probabilities, and we have yet to define those. This will be taken up in the next section.

We are making the simplifying assumption that actions occur instantaneously, and so their dates are time points rather than intervals. If a state of affairs neither predates $A$ nor postdates $A$ then its date must be an interval (possibly with gaps) with the date of $A$ lying within the interval. I assume that such a state of affairs can be split into a "first part" predating $A$ and a "second part" postdating $A$, and then the state of affairs can be represented as the conjunction of these two parts. This has the consequence that $\text{C-PROB}_A(O)$ is defined for all states of affairs $O$.[9]

We can construct a version of causal decision theory by defining expected values in terms of C-PROB:

$$\mathbf{EV}(A) = \sum_{O\in\mathbf{O}} \mathrm{U}(O)\cdot\text{C-PROB}_A(O).$$

I will call this *T-causal decision theory* because of its reliance on temporal ordering rather than causal dependence.

## 5. Conditional Causal Probability

Now let us turn to the question of how to define conditional causal probabilities. The standard definition would be:

$$\text{C-PROB}_A(O/P) = \text{C-PROB}_A(O\&P)/\text{C-PROB}_A(P).$$

We can adopt this as our definition in all cases except when $P$ predates $A$. However, in that case we defined $\text{C-PROB}_A(O\&P)$ in terms of $\text{C-PROB}_A(O/P)$. Thus when $P$ predates $A$, we need a different defintion for conditional causal probability. Two possibilities may occur to us regarding how to define these conditional causal probabilities:

(1) $$\text{C-PROB}_A(O/P) = \sum_{W\in\mathbf{W}} \text{PROB}(W)\cdot\text{PROB}(O/A\&W\&P).$$

[9] This is the only place we need the assumption that actions occur instantaneously. As long as the time interval over which the action occurs is disjoint from that of $O$, there is no problem defining $\text{C-PROB}_A(O)$.

(2) $\text{C-PROB}_A(O/P) = \sum_{W \in \mathbf{W}} \text{PROB}(W/P) \cdot \text{PROB}(O/A\&W\&P)$.

The issue is whether we should conditionalize the probabilities of the backgrounds on $P$. To see that (2) is the correct definition, consider a case in which a background $B$ is made more probable by $P$, but *given* $B$, $P$ has no effect on the probability that the outcome will result from doing $A$. For example, some people are more likely to get lung cancer from smoking than others. Suppose there is a genetic factor $G$ involved. So $\text{PROB}(LC/S\&G) > \text{PROB}(LC/S\&{\sim}G)$. Let $D$ be a diagnostic result that makes it more probable that you have $G$. Then it ought to be the case that $\text{C-PROB}_S(LC/D) > \text{C-PROB}_S(LC/{\sim}D)$. Presumably, *given* $G$ or $\sim G$, adding information about $D$ should not affect the probability of given lung cancer given that you smoke. That is, $D$ is only relevant because it gives us information about $G$. So $\text{PROB}(LC/S\&G\&D) = \text{PROB}(LC/S\&G)$, $\text{PROB}(LC/S\&{\sim}G\&D) = \text{PROB}(LC/S\&{\sim}G)$, $\text{PROB}(LC/S\&G\&{\sim}D) = \text{PROB}(LC/S\&G)$, and $\text{PROB}(LC/S\&{\sim}G\&{\sim}D) = \text{PROB}(LC/S\&{\sim}G)$. Then on proposal (1) we would have

$$\text{C-PROB}_S(LC/D)$$

$$= \text{PROB}(G) \cdot \text{PROB}(LC/S\&G\&D) + \text{PROB}({\sim}G) \cdot \text{PROB}(LC/S\&{\sim}G\&D)$$

$$= \text{PROB}(G) \cdot \text{PROB}(LC/S\&G) + \text{PROB}({\sim}G) \cdot \text{PROB}(LC/S\&{\sim}G)$$

$$\text{C-PROB}_S(LC/{\sim}D)$$

$$= \text{PROB}(G) \cdot \text{PROB}(LC/S\&G\&{\sim}D) + \text{PROB}({\sim}G) \cdot \text{PROB}(LC/S\&{\sim}G\&{\sim}D)$$

$$= \text{PROB}(G) \cdot \text{PROB}(LC/S\&G) + \text{PROB}({\sim}G) \cdot \text{PROB}(LC/S\&{\sim}G)$$

So proposal (1) yields the incorrect result that $\text{C-PROB}_S(LC/D) = \text{C-PROB}_S(LC/{\sim}D)$. On the other hand, according to proposal (2) we would have

$$\text{C-PROB}_S(LC/D)$$

$$= \text{PROB}(G/D) \cdot \text{PROB}(LC/S\&G\&D) + \text{PROB}({\sim}G/D) \cdot \text{PROB}(LC/S\&{\sim}G\&D)$$

$$= \text{PROB}(G/D) \cdot \text{PROB}(LC/S\&G) + \text{PROB}({\sim}G/D) \cdot \text{PROB}(LC/S\&{\sim}G)$$

$$\text{C-PROB}_S(LC/{\sim}D)$$

$$= \text{PROB}(G/{\sim}D) \cdot \text{PROB}(LC/S\&G\&{\sim}D)$$

$$+ \text{PROB}({\sim}G/{\sim}D) \cdot \text{PROB}(LC/S\&{\sim}G\&{\sim}D)$$

$$= \text{PROB}(G/{\sim}D) \cdot \text{PROB}(LC/S\&G) + \text{PROB}({\sim}G/{\sim}D) \cdot \text{PROB}(LC/S\&{\sim}G)$$

We assumed that $\textbf{PROB}(LC/S\&G) > \textbf{PROB}(LC/S\&{\sim}G)$, and $\textbf{PROB}(G/D) > \textbf{PROB}(G/{\sim}D)$, so we get the correct result that $\textbf{C-PROB}_S(LC/D) > \textbf{C-PROB}_S(LC/{\sim}D)$.

Accordingly, I will define conditional causal probabilities as in (2). When $P$ predates $A$, our official definition is (2). When $P$ predates $A$, our official definition is:

$$\textbf{C-PROB}_A(O/P) = \sum_{W \in \mathbf{W}} \textbf{PROB}(W/P) \cdot \textbf{PROB}(O/A\&W\&P).$$

This has the consequence that $P$ functions informationally while $A$ functions causally. That is, $P$ can have backward ramifications, influencing the probability of backgrounds, but $A$ can only influence the probabilities of future events.

## 6. $\textbf{C-PROB}_A$ and $\textbf{K-PROB}_A$

T-causal decision theory handles the counterexamples to classical decision theory in essentially the same way other causal decision theories do, but it defines causal probability without appeal to a vague undefined concept of causal dependence. Without a clear definition of causal dependence, it seems to me that the appeal to the evolution of scenarios in temporal order is a more obvious diagnosis of the counterexamples than is the appeal to causal dependence. It resolves the counterexamples in an intuitively congenial way, without appealing to anything more problematic than temporal ordering and the fact that causation propagates forwards in time.

Without an analysis, the concept of causal dependence is sufficiently unclear that the behavior of a concept of causal probability defined in terms of it is not clear. However, as I will now show, if we confine our attention to point-dated actions and singular states of affairs postdating $A$ and make two plausible assumptions about causal dependence, it follows that T-causal decision theory is equivalent to Skyrms' theory (on the narrow construal). Thus we get an equivalent theory without appealing to the vague notion of causal dependency.

### *6.1 Causation and Time*

The first assumption we need is that if $P$ predates $A$, then $P$ is causally independent of $A$. Plausibly, this follows from another principle, viz., that if $P$ causes $Q$ then $P$ predates $Q$. Consider then the principle that if $P$ causes $Q$ then $P$ predates $Q$. I take it that, in normal contexts, this is obvious and uncontroversial. I have always been puzzled by why some philosophers think it is a problem for a theory of causation if it does not allow backwards causation. Backwards causation seems to me to be a logical absurdity. This is surely the commonsense view. To maintain otherwise requires an argument.

Is there an argument for the possibility of backwards causation? Some readers may think of the possibility of time travel here. If time travel is possible, then I might cause earlier events to occur by traveling backwards in time. This has always seemed logically puzzling, which is the reason stories of time travel intrigue their readers. My own view is that time travel is a logical impossibility. One can argue for this conclusion by appealing to the so-called paradoxes of time travel. For example, I might travel backwards in time and kill my grandfather. But then I would not have existed, so I could not have done that. This generates a contradiction. Some philosophers (for example, Lewis 1976) are content to assert that this is not paradoxical—it merely shows that if I did travel back in time I could not kill my grandfather. But that is not a satisfying solution to the paradox. If I have traveled back to the time of my grandfather, we need some explanation for why I would not have the same causal powers as everyone else. It is not sufficient to say that I just couldn't (with no explanation).

Although the principle that causes predate their effects is an intuitive one, we need not rest our case on that. We can also give arguments for it. The simplest argument is that this must be assumed to make sense of the direction of causation. Consider the relationship between causation and nomic implication. The simplest and clearest cases of causation arise when laws of nature entail that whenever *P* occurs, it will be followed by *Q*. In that case at least, *P's* being true causes *Q* to be true later. Consider a world in which it is a law of nature that some state *P* oscillates with the regular interval $\Delta$. So *P's* being true at *t* nomically implies that *P* is false at $t+\Delta$, and that *P* is true again at $t+2\Delta$. Then it follows that *P's* being true again at $t+2\Delta$ nomically implies that *P* was true at *t*. We want to say that *P's* being true at *t* causes *P* to be true at $t+2\Delta$, but deny that *P's* being true at $t+2\Delta$ causes *P* to be true at *t*. Why? As far as the laws of nature are concerned, their relationship is perfectly symmetric. The only difference between them is that one comes before the other. We take the earlier one to cause the later one, but not conversely. It seems to me that the only way to sort out the direction of causation is to make it parasitic on the direction of time. But then it is a necessary truth that causes predate their effects.

A deeper argument for the principle that causes must predate their effects turns upon the way we reason about causes and effects. This is tied up with the extensive literature on the Frame Problem in AI. The Frame Problem is, basically, the problem of understanding how causal reasoning works.[10] It is now generally agreed that the solution to the Frame Problem turns upon two principles. These can be illustrated by looking at an example that is known in the AI literature as "The Yale Shooting Problem", and is due to Steve Hanks and Drew McDermott (1986,1987). Suppose Smith loads his gun, points it at Jones, and pulls the trigger, and suppose (a bit simplistically) that we have a causal law to the effect that if a loaded gun is pointed at someone and the trigger pulled, they will shortly be dead. Then we want to conclude that Jones will die. There are, however, two problems for drawing this conclusion. First, time passed between when Smith loaded the gun and

[10] See Pollock (1998a) for a more sustained discussion of this claim.

when he pulled the trigger. How do we know that the gun is still loaded? We ordinarily assume that if a gun is loaded at one time then it is still loaded a short while later unless we have some reason for thinking otherwise. A bit more precisely, we assume a principle of *Temporal Projection* that can be stated roughly as follows:

> If $P$ is "temporally projectible", then $P$'s being true at time $t_0$ gives us a defeasible reason for expecting $P$ to be true at a later time $t$, the strength of the reason being a monotonic decreasing function of the time difference $(t - t_0)$.[11]

This is also known as the "commonsense law of inertia". The temporal projectibility constraint reflects the fact that this principle does not hold for some choices of $P$. For instance, its being three o'clock at one time does not give us a reason for expecting it to be three o'clock ten minutes later.

This principle licenses a defeasible inference to the conclusion that the gun is still loaded when the trigger is pulled. Then from our causal law, we can infer that Jones will die. But this does not yet solve the problem. The difficulty is that Temporal Projection also licenses a defeasible inference to conclusion that because Jones was originally alive, he will still be alive after the trigger is pulled. So we have defeasible arguments for both the conclusion that Jones will die and the conclusion that he will not. We need a further principle that explains why we prefer the former conclusion to the latter.

It is now generally agreed that the solution to the Yale Shooting Problem is that in causal reasoning we think of the world as unfolding temporally. By temporal projection, we expect things to remain unchanged until something happens to force them to change. Thus we expect the gun to remain loaded because nothing has happened to cause it to become unloaded. However, once the gun is fired, something has happened to change the state of Jones' health. This is captured by the principle that when two temporal projections conflict causally, as in the Yale Shooting Problem, we give precedence to the earlier one. Thus we assume that gun will remain loaded until fired, but not that Jones willl remain alive. Formally, this is captured by the following principle of "causal undercutting", where $A$ is an action:

> If $t_0 < t < t^*$ and $t_{00} < t^*$, "$P$-at-$t_0$ and $A$-at-$t$ and [($P$-at-$t$ & $A$-at-$t$) is causally sufficient for $Q$ at $t^*$]" is an undercutting defeater for the inference by temporal projection from "~$Q$-at-$t_{00}$" to "~$Q$-at-$t^*$".

I showed in my (1998a) that with the help of temporal projection and causal undercutting, we can conclude that Jones will die. This depend upon details about how defeasible reasoning works, so I will not try to reproduce the argument here, but the interested reader can look at the original article.[12]

---

[11] See Pollock (1998a) for further discussion of this principle.

[12] Or see Pollock (2007a) for an exposition of my account of causal reasoning.

These two principles seem to capture the logic of our ordinary causal reasoning, but they only do so provided causes must predate their effects. If we suppose that a cause can have a prior effect, then we could not give precedence to earlier temporal projections over later ones. But then causal reasoning would be impossible. This would have the effect of making our causal reasoning unworkable. This, I think, is a deeper reason why our concept of causation requires causes to predate effects.

In light of considerations like these, I am convinced that our ordinary concept of causation requires that causes predate their effects. So I assume that if *P* predates *A* then *P* is causally independent of *A*.

### *6.2 The Markov Condition*

My first assumption has the consequence that the elements of a background *W* in $\mathbf{W}$ are causally independent of *A*. If $K \in \mathbf{K}$ then because *K* is a complete specification of causally independent states of affairs, it follows that there is a background *W*(*K*) in $\mathbf{W}$ such that $W(K) \subseteq K$. *K* will also contain many states of affairs postdating *A*. Most of them will be statistically independent of *A* in the sense that, if $K_0$ is the set of them, then $\textbf{PROB}(K_0/A\&(K-K_0)) = \textbf{PROB}(K_0/(K-K_0))$. It then follows as in theorem 3 below that omitting them from *K* will not affect the calculation of $\textbf{K-PROB}_A(O)$. For the remaining elements of *K* that postdate *A*, *A* is statistically relevant to them but they are causally independent of *A*. My second assumption is that this is only possible if either *A* or those remaining elements of *K* have a common cause. This is a familiar assumption in statistics (Kiiveri and Speed 1982), and is sometimes called "the Markov condition". To be causally relevant to *A*, that common cause must lie in the part of *K* that predates *A*, that is, it must lie in *W*(*K*). So the precise assumption I will make is that $\textbf{PROB}(O/A\&K) = \textbf{PROB}(O/A\&W(K))$. It is to be emphasized that this is a considerable precization of a rather vague assumption about causal relevance. Let $\mathbf{K}^* = \{K - W(K) \mid K \in \mathbf{K}\}$. For any $W \in \mathbf{W}$, *W* is equivalent to the disjunction of all $W\&K^*$ for $K^* \in \mathbf{K}^*$. So

$$\sum_{K^* \in \mathbf{K}^*} \textbf{PROB}(W\&K^*) = \textbf{PROB}(W).$$

It then follows that $\textbf{K-PROB}_A(O) = \textbf{C-PROB}_A(O)$.

The proof is as follows:

$$\begin{aligned}
\textbf{K-PROB}_A(O) &= \sum_{K\in\mathbf{K}} \textbf{PROB}(O/A\&K)\cdot\textbf{PROB}(K) \\
&= \sum_{K\in\mathbf{K}} \textbf{PROB}(O/A\&W(K))\cdot\textbf{PROB}(K) \\
&= \sum_{W\in\mathbf{W}}\sum_{K^*\in\mathbf{K}^*} \textbf{PROB}(O/A\&W(K^*\&W))\cdot\textbf{PROB}(K^*\&W) \\
&= \sum_{W\in\mathbf{W}}\sum_{K^*\in\mathbf{K}^*} \textbf{PROB}(O/A\&W)\cdot\textbf{PROB}(K^*\&W) \\
&= \sum_{W\in\mathbf{W}} \textbf{PROB}(O/A\&W)\cdot\sum_{K^*\in\mathbf{K}^*} \textbf{PROB}(K^*\&W)
\end{aligned}$$

$$= \sum_{W \in \mathbf{W}} \mathrm{PROB}(O/A \& W) \cdot \mathrm{PROB}(W)$$
$$= \text{C-PROB}_A(O).$$

Thus if we make these two assumptions about causal dependence, $\text{C-PROB}_A(O) = \text{K-PROB}_A(O)$, and hence T-causal decision theory is equivalent to Skyrms' theory. However, T-causal decision theory has the advantage that causal probability is defined without reference to causal dependence. On the other hand, if it is proposed that causal dependence should be understood in a way that makes the Markov assumption be false, then I suggest that $\text{C-PROB}_A$ is the more obvious choice for causal probability and handles the counterexamples to classical decision theory more tidily than do other causal decision theories.

My discussion of probabilistic causation raises the question of in what sense my causal probabilities are causal. But nothing turns upon whether they really are. For the sake of argument, I have been granting that there is some kind of probabilistic causal connection lurking in the background here, but it is not at all clear what that amounts to or whether there is any way to make sense of such a probabilistic causal connection. For present purposes, that is not important. What we want are probabilities that enable us to reason in the right way about decisions. Whether those probabilities are causal is irrelevant.

### *6.3 Causal Decision Theory and Evidential Decision Theory*

As noted in section 1, several authors have argued that there are ways of reconstruing evidential decision theory so that it does not lead to the objectionable conclusions that motivate causal decision theory (Eells 1981; Jeffrey 1983; Price 1991). I am unconvinced that these attempts are successful, but my objective here is not to argue that evidential decision theory is wrong. My concern is to argue that causal decision theory, as I formulated it, is correct. If it turns out that evidential decision theory is able to make the same choices by making the reasoning more complex, that is interesting, but it does not show that my theory is wrong. Once the intuitive distinction between causal decision theory and evidential decision theory is pointed out, it is hard to see why anyone would want to defend evidential decision theory except as a curiosity. The intuitions giving rise to decision theory concern the probabilities that our actions will have various effects, and that is what causal decision theory aims to make precise. Given this, evidential decision theory can only be saved by showing that it agrees with causal decision theory, but showing that is in no way a criticism of causal decision theory.

## 7. Computing Causal Probabilities

If causal probability is to be useful, there must be an efficient way of computing it. If we had to compute $\text{C-PROB}_A(O)$ by actually performing the summation (or integration) involved in the definition, the task would be formidable. Fortunately, this computation can be simplified considerably. Unfortunately, the details are technically complex, so the reader who is only interested in a philosophical overview of causal probability may want to skip to section 9.

The fundamental idea behind this definition of causal probability is that in computing how likely an outcome is to result from an action, we want to propagate changes forward in time rather than backward. A useful way of conceptualizing this is to think of the world as described by different *scenarios*, each consisting of some initial *A*-world-state being true, followed by the action, followed by an outcome. The scenarios can be diagrammed in the form of a tree, as in figure 1. $\mathbf{C\text{-}PROB}_A(O)$ should then be of the probability of the disjunction of the scenarios that terminate with *O*. If there were just one such scenario, the probability associated with it would be $\mathbf{PROB}(W)\cdot\mathbf{PROB}(O/A\&W)$, and that would be the value of $\mathbf{C\text{-}PROB}_A(O)$. The probability associated with the scenario results from propagating the probabilities of changes forward in time. We can identify $\mathbf{PROB}(W)\cdot\mathbf{PROB}(O/A\&W)$ with $\mathbf{C\text{-}PROB}_A(O\&W)$ because (1) $\mathbf{C\text{-}PROB}_A(O\&W) = \mathbf{C\text{-}PROB}_A(W)\cdot\mathbf{C\text{-}PROB}_A(O/W)$, (2) $\mathbf{C\text{-}PROB}_A(W) = \mathbf{PROB}(W)$ because *W* predates *A* and so cannot be affected by it, and (3) $\mathbf{C\text{-}PROB}_A(O/W) = \mathbf{PROB}(O/A\&W)$ because *W* includes everything that is relevant to whether *O* will result from performing *A*.

Suppose there are *n* possible scenarios, associated with the initial *A*-world-states $W_1,\ldots,W_n$. Then the disjunction of the scenarios is $(W_1\&A\&O) \vee\ldots\vee (W_n\&A\&O)$, which is equivalent to $(W_1 \vee\ldots\vee W_n)\&A\&O$. We can think of this as a single scenario with a disjunctive background state and identify $\mathbf{C\text{-}PROB}_A(O)$ with the probability of this scenario. The different $W_i$'s are logically disjoint, so the probability associated with this scenario can be computed as follows:

$$\mathbf{C\text{-}PROB}_A(O) = \mathbf{C\text{-}PROB}_A((W_1 \vee\ldots\vee W_n)\&A\&O)$$

$$= \mathbf{C\text{-}PROB}_A(W_1 \vee\ldots\vee W_n)\cdot\mathbf{C\text{-}PROB}_A(O/W_1 \vee\ldots\vee W_n)$$

$$= \mathbf{C\text{-}PROB}_A(W_1 \vee\ldots\vee W_n) \cdot[\mathbf{C\text{-}PROB}_A(O/W_1)\cdot\mathbf{C\text{-}PROB}_A(W_1/W_1 \vee\ldots\vee W_n) + \ldots + \mathbf{C\text{-}PROB}_A(O/W_n)\cdot\mathbf{C\text{-}PROB}_A(W_n/W_1 \vee\ldots\vee W_n)]$$

$$= \mathbf{C\text{-}PROB}_A(O/W_1)\cdot\mathbf{C\text{-}PROB}_A(W_1) + \ldots + \mathbf{C\text{-}PROB}_A(O/W_n)\cdot\mathbf{C\text{-}PROB}_A(W_n)$$

$$= \mathbf{PROB}(W_1)\cdot\mathbf{C\text{-}PROB}_A(O/W_1) + \ldots + \mathbf{PROB}(W_n)\cdot\mathbf{C\text{-}PROB}_A(O/W_n)$$

$$= \mathbf{PROB}(W_1)\cdot\mathbf{PROB}(O/A\&W_1) + \ldots + \mathbf{PROB}(W_n)\cdot\mathbf{PROB}(O/A\&W_n).$$

In other words, $\mathbf{C\text{-}PROB}_A(O)$ is the probability associated with the disjunctive scenario, and that in turn is the sum of the probabilities associated with the individual scenarios.

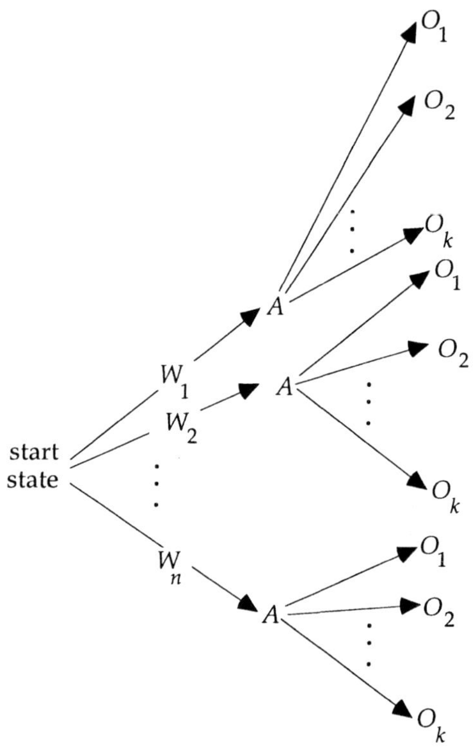


Figure 8.1. Scenarios evolving with the passage of time

There is a similar way of conceptualizing conditional causal probabilities. We defined:

$$\textbf{C-PROB}_A(O/P) = \sum_{W\in\mathbf{W}} \textbf{PROB}(W/P)\cdot\textbf{PROB}(O/A\&W\&P).$$

This is what we get from figure 8.1 if we replace the start state by the supposition *P*.

Recall that $\mathbf{W}$ is the set of initial *A*-world-states and $\mathbf{C}$ is the set of "constituents" of initial *A*-world-states. Let us say that a subset *S* of $\mathbf{C}$ *shadows A* with respect to *O* iff (1) *S* is nomically consistent with *A*, (2) for every $W\in\mathbf{W}$ and any $S^{**}$, if $S \subseteq S^{**} \subseteq W$ then $\textbf{PROB}(O/A\&S^{**}) = \textbf{PROB}(O/A\&S)$, and (3) there is no proper subset $S^*$ of *S* such that for every $W\in\mathbf{W}$ and any $S^{**}$, if $S^* \subseteq S^{**} \subseteq W$ then $\textbf{PROB}(O/A\&S^{**}) = \textbf{PROB}(O/A\&S)$. The shadows are minimal descriptions of all aspects of the initial *A*-world-state relevant to the evaluation of the probability of *O*. Let $\mathbf{S}$ be the set of all shadows. Shadows can be constructed by starting from members of $\mathbf{W}$ and then removing elements that do not affect the probability of *O*.

It follows that every initial *A*-world-state *W* contains a shadow *S* such that **PROB**(*O*/*A*&*W*) = **PROB**(*O*/*A*&*S*).

Let $\mathbf{C}^*$ be the set of all members of $\mathbf{C}$ occurring in one or more of the shadows. Define a *background* to be a maximal subset of $\mathbf{C}^*$ nomically consistent with *A*. Let $\mathbf{B}$ be the set of all backgrounds. The backgrounds form a partition. That is, they are pairwise logically disjoint and the disjunction of all of them is a (nomically) necessary truth. Now suppose $B \in \mathbf{B}$ and $W \in \mathbf{W}$ and $B \subseteq W$. *W* contains a shadow *S* such that **PROB**(*O*/*A*&*W*) = **PROB**(*O*/*A*&*S*), and the shadow consists of members of $\mathbf{C}^*$, so $S \subseteq B$, and hence by the definition of "shadow", **PROB**(*O*/*A*&*B*) = **PROB**(*O*/*A*&*S*). Thus **PROB**(*O*/*A*&*W*) = **PROB**(*O*/*A*&*B*). Now we can prove a central theorem in the theory of causal probability:

**Theorem 1:** $$\textbf{C-PROB}_A(O) = \sum_{B \in \mathbf{B}} \textbf{PROB}(B) \cdot \textbf{PROB}(O/A\&B).$$

Proof: For $B \in \mathbf{B}$, let $W(B) = \{W | W \in \mathbf{W} \ \&\ B \subseteq W\}$. *B* is equivalent to the disjunction of the members of *W*(*B*). Then

$$\textbf{C-PROB}_A(O) = \sum_{W \in \mathbf{W}} \textbf{PROB}(W) \cdot \textbf{PROB}(O/A\&W)$$

$$= \sum_{B \in \mathbf{B}} \sum_{W \in W(B)} \textbf{PROB}(W) \cdot \textbf{PROB}(O/A\&W)$$

$$= \sum_{B \in \mathbf{B}} \sum_{W \in W(B)} \textbf{PROB}(W) \cdot \textbf{PROB}(O/A\&B)$$

$$= \sum_{B \in \mathbf{B}} \textbf{PROB}(O/A\&B) \cdot \sum_{W \in W(B)} \textbf{PROB}(W)$$

$$= \sum_{B \in \mathbf{B}} \textbf{PROB}(O/A\&B) \cdot \textbf{PROB}(B). \ \blacksquare$$

$\mathbf{W}$ is immense (in fact, infinite), but $\mathbf{B}$ may be very small. In the smoking gene example, if we suppose that the only part of an initial *S*-world-state that makes any difference to the probability of getting cancer is *G* or ~*G*, it follows that $\mathbf{S} = \{\{G\},\{\sim G\}\}$, and so $\mathbf{B} = \{\{G\},\{\sim G\}\}$, and hence

$$\textbf{C-PROB}_S(cancer)$$

$$= \textbf{PROB}(G) \cdot \textbf{PROB}(cancer/S\&G) + \textbf{PROB}(\sim G) \cdot \textbf{PROB}(cancer/S\&\sim G).$$

Of course, realistically, other elements of *S*-world-states will also be statistically relevant, for example, whether one's parents had the smoking gene. However, the effect of one's parents having the smoking gene is "screened off" by knowing whether one has the gene oneself. That is, if you know whether you have the smoking gene, the additional knowledge of whether your parents had it does not effect the probability of getting cancer. So the set of shadows, and hence the set of backgrounds, remains unchanged.

Normally, shadows will be more numerous than in the smoking gene example.

However, the shadows may not all be relevant. The need for causal probabilities only arises when the action is statistically relevant to some of the backgrounds. If the backgrounds are all statistically independent of the action, then the causal probability is the same as the mixed physical/epistemic probability:

**Theorem 2:** If for each $B \in \mathbf{B}$, **PROB**($B/A$) = **PROB**($B$), then **C-PROB**$_A$($O$) = **PROB**($O/A$).

More generally, the action may be statistically relevant to just a few constituents of the backgrounds. Then we can often make use of the following theorem:

**Theorem 3:** If $\mathbf{C}_0 \subseteq \mathbf{C}^*$, let $\mathbf{B}_0$ be the set of all maximal subsets of $\mathbf{C}_0$ nomically consistent with $A$, and let $\mathbf{B}^*$ be the set of all maximal subsets of $\mathbf{C}^* - \mathbf{C}_0$ nomically consistent with $A$. If for every $B_0 \in \mathbf{B}_0$ and $B^* \in \mathbf{B}^*$, **PROB**($B^*/B_0$&$A$) = **PROB**($B^*/B_0$), then

$$\textbf{C-PROB}_A(O) = \sum_{B \in \mathbf{B}_0} \textbf{PROB}(B_0)\cdot\textbf{PROB}(O/A\&B_0).$$

Proof: The backgrounds $B$ are just the conjunctions (unions) of a $B_0 \in \mathbf{B}_0$ and a $B^* \in \mathbf{B}^*$, and the disjunction of the members of $\mathbf{B}^*$ is a necessary truth, so

$$\textbf{C-PROB}_A(O) = \sum_{B \in \mathbf{B}} \textbf{PROB}(B)\cdot\textbf{PROB}(O/A\&B)$$

$$= \sum_{B_0 \in \mathbf{B}_0} \sum_{B^* \in \mathbf{B}^*} \textbf{PROB}(B_0\&B^*)\cdot\textbf{PROB}(O/A\&B_0\&B^*)$$

$$= \sum_{B_0 \in \mathbf{B}_0} \sum_{B^* \in \mathbf{B}^*} \textbf{PROB}(B_0)\cdot\textbf{PROB}(B^*/B_0)\cdot\textbf{PROB}(O/A\&B_0\&B^*)$$

$$= \sum_{B_0 \in \mathbf{B}_0} \textbf{PROB}(B_0)\cdot\sum_{B^* \in \mathbf{B}^*} \textbf{PROB}(B^*/B_0)\cdot\textbf{PROB}(O/A\&B_0\&B^*)$$

$$= \sum_{B_0 \in \mathbf{B}_0} \textbf{PROB}(B_0)\cdot\sum_{B^* \in \mathbf{B}^*} \textbf{PROB}(B^*/B_0\&A)\cdot\textbf{PROB}(O/A\&B_0\&B^*)$$

$$= \sum_{B \in \mathbf{B}_0} \textbf{PROB}(B_0)\cdot\textbf{PROB}(O/A\&B_0). \blacksquare$$

So if there is a subset $\mathbf{C}_0$ of constituents of backgrounds relative to which all other combinations of constituents are statistically independent of $A$, then we can compute causal probabilities by making reference only to backgrounds built out of the members of $\mathbf{C}_0$. For example, suppose there are two constituents of backgrounds that are statistically relevant to getting cancer — having the smoking gene, and having been raised on a nuclear waste dump ($N$). Then $\mathbf{B}$ = {{$G,N$},{$G,\sim N$},{$\sim N,G$},{$\sim N,\sim G$}}. However, $S$ is not statistically relevant to whether one was raised on a nuclear waste dump, even given that one does or does not have the smoking gene:

**PROB**($N/S\&G$) = **PROB**($N/G$)

**PROB**($N/S\&\sim G$) = **PROB**($N/\sim G$)

$\mathbf{PROB}(\sim N/S\&G) = \mathbf{PROB}(\sim N/G)$

$\mathbf{PROB}(\sim N/S\&\sim G) = \mathbf{PROB}(\sim N/\sim G)$

So we can let $\mathbf{C}_0 = \{\{G\},\{\sim G\}\}$, and once more compute $\mathbf{C\text{-}PROB}_S(cancer)$ by reference to the small set of backgrounds $\mathbf{B}_0 = \{\{G\},\{\sim G\}\}$.

The upshot of these results is that causal probabilities will usually be computable by performing manageably small sums. In cases in which actions are statistically relevant to their backgrounds, **C-PROB**'s may be significantly easier to compute than **PROB**'s. **C-PROB**'s can be computed recursively by propagating probabilities forwards through scenarios. But if a later state can affect the **PROB** of an earlier state, then **PROB**'s cannot similarly be computed recursively. For practical purposes, **C-PROB**'s are simpler than **PROB**'s. This suggests that instead of expressing theorem 2 by saying that causal probabilities usually behave classically, it might be better to say that classical probabilities usually behave causally.

## 8. Computing Conditional Causal Probabilities

As in the case of non-conditional causal probabilities, the computation of conditional causal probabilities can also be simplified by recasting it in terms of backgrounds. Just as in theorem 1, we have:

**Theorem 4:** If $P$ predates $A$ then where $\mathbf{B}$ is the set of backgrounds for $O$ relative to $A$ that are consistent with $P$:

$$\mathbf{C\text{-}PROB}_A(O/P) = \sum_{B\in\mathbf{B}} \mathbf{PROB}(B/P)\cdot\mathbf{PROB}(O/A\&B\&P).$$

The proof is analogous to that of theorem 1.

From theorem 4 we get a simple theorem that will be repeatedly useful:

**Theorem 5:** If $B$ is a background for $A$ relative to $O$ and $P$ predates $A$ then $\mathbf{C\text{-}PROB}_A(O/B\&P) = \mathbf{PROB}(O/A\&B\&P)$.

Proof: $\mathbf{PROB}(B/B\&P) = 1$, and for any other background $B^*$, $\mathbf{PROB}(B^*/B\&P) = 0$. ■

We also get theorems analogous to theorems 2 and 3:

**Theorem 6:** If $P$ predates $A$ and for every background $B$ for $A$ relative to $O$ that is consistent with $P$, $\mathbf{PROB}(B/A\&P) = \mathbf{PROB}(B/P)$, then $\mathbf{C\text{-}PROB}_A(O/P) = \mathbf{PROB}(O/A\&P)$.

**Theorem 7:** If $\mathbf{C}_0 \subseteq \mathbf{C}^*$, let $\mathbf{B}_0$ be the set of all maximal subsets of $\mathbf{C}_0$ nomically consistent with $A\&P$, and let $\mathbf{B}^*$ be the set of all maximal subsets of $\mathbf{C}^* - \mathbf{C}_0$ nomically consistent with $A\&P$. If for every $B_0\in\mathbf{B}_0$ and $B^*\in\mathbf{B}^*$, $\mathbf{PROB}(B^*/B_0\&A\&P) = \mathbf{PROB}(B^*/B_0\&P)$, then

$$\textbf{C-PROB}_A(O/P) = \sum_{B\in\mathbf{B}_0} \textbf{PROB}(B_0/P)\cdot\textbf{PROB}(O/A\&B_0\&P).$$

By virtue of these theorems, in computing conditional causal probabilities we can usually restrict our attention to very small backgrounds.

Just as for nonconditional probabilities, the conditional causal probabilities of states predating *A* behave classically:

**Theorem 8:** If *P* and *Q* predate *A*, $\textbf{C-PROB}_A(Q/P) = \textbf{PROB}(Q/P)$.

Proof: $\textbf{C-PROB}_A(Q/P) = \textbf{C-PROB}_A(Q\&P)/\textbf{C-PROB}_A(P) = \textbf{PROB}(Q\&P)/\textbf{PROB}(P) = \textbf{PROB}(Q/P)$. ■

We say that a case is *classical* iff for every background *B* for *A* relative to *O*, $\textbf{PROB}(B/A\&P) = \textbf{PROB}(B/P)$. By theorems 2 and 6, in classical cases $\textbf{C-PROB}_A(O) = \textbf{PROB}(O/A)$ and $\textbf{C-PROB}_A(O/P) = \textbf{PROB}(O/A\&P)$.

## 9. Simplifying the Computation Defeasibly

Decision theory is a theory about how cognizers should, rationally, direct their activities. Causal decision theory tells them to use causal probabilities in their deliberations, and to do that they must have beliefs about causal probabilities. Sections 7 and 8 show that there are, in principle, feasible ways of computing these probabilities. However, causal probabilities are defined in terms of mixed physical/epistemic probabilities, which are in turn inferred by direct inference from nomic probabilities. To apply the definitions directly a cognizer would have to know all the relevant nomic probabilities and compute all the relevant mixed physical/epistemic probabilities. Real cognizers will fall far short of this ideal. They have limited knowledge of nomic probabilities, and correspondingly limited access to the values of mixed physical/epistemic probabilities and causal probabilities. Despite this, these three kinds of probabilities are useful because when cognizers lack direct knowledge of them, they can still estimate them defeasibly using classical and nonclassical direct inference. Direct inference applies directly to mixed physical/epistemic probabilities and nomic probabilities, but as I will show it also supports defeasible inferences regarding causal probabilities.

It is useful to remember that the principles of classical and nonclassical direct inference are theorems of the theory of nomic probability—not primitive assumptions. They follow from the principle (SS) of the statistical syllogism and the calculus of nomic probabilities. Classical direct inference tells us how to infer the values of mixed physical/epistemic probabilities from the values of associated nomic probabilities. The core principle for computing non-conditional mixed physical/epistemic probabilities was (CDI), discussed in chapter 7. The analogous principle for conditional mixed physical/epistemic probability is:

(CDI*) If $A$ is projectible with respect to $B$ and $C$ then "$\mathbf{K}(Q \leftrightarrow Cc)$ & $\mathbf{K}[Q \rightarrow (P \leftrightarrow Ac)]$ & $\mathbf{K}Bc$ & $\mathbf{prob}(Ax/Bx \,\&\, Cx) = r$" is a defeasible reason for "$\mathrm{PROB}(P/Q) = r$".

In effect, this principle tells us that if the cognizer know that $[Q \rightarrow (P \leftrightarrow Ac)]$ and $(Q \leftrightarrow Cc)$, then he can identify $\mathrm{PROB}(P/Q)$ with $\mathrm{PROB}(Ac/Cc)$, and if the cognizer is knows that $Bc$ he can defeasibly expect $\mathrm{PROB}(Ac/Cc)$ to be the same as $\mathbf{prob}(Ax/Bx \,\&\, Cx)$. A corollary of this is that it is defeasibly reasonable to expect any further information we might acquire about $c$ to be irrelevant to the value of $\mathrm{PROB}(Ac/Cc)$.

Nonclassical direct inference has a similar flavor, telling us that it is defeasibly reasonable to expect further projectible properties $C$ to be statistically irrelevant to the nomic probability $\mathbf{prob}(Ax/Bx)$:

(DI) If $A$ is projectible with respect to $B$ then "$\mathbf{prob}(Ax/Bx) = r$" is a defeasible reason for "$\mathbf{prob}(Ax/Bx \,\&\, Cx) = r$".

From these two principles, we can derive a defeasible presumption of statistical irrelevance for mixed physical/epistemic probabilities:

(IR) For any $P,Q,R$, it is defeasibly reasonable to expect that $\mathrm{PROB}(P/Q\&R) = \mathrm{PROB}(P/Q)$.

This conclusion is forthcoming from the preceding principles of classical and non-classical direct inference. If the cognizer is justified in believing $(P \leftrightarrow Ac)$, $(Q \leftrightarrow Cc)$, $(R \leftrightarrow Dc)$, and $Bc$, it is defeasibly reasonable for him to conclude that $\mathrm{PROB}(P/Q) = \mathbf{prob}(Ax/Bx \,\&\, Cx)$ and to conclude that $\mathrm{PROB}(P/Q\&R) = \mathbf{prob}(Ax/Bx \,\&\, Cx \,\&\, Dx)$. But by non-classical direct inference, it is also defeasibly reasonable to expect that $\mathbf{prob}(Ax/Bx \,\&\, Cx \,\&\, Dx) = \mathbf{prob}(Ax/Bx \,\&\, Cx)$, so it is defeasibly reasonable to expect that $\mathrm{PROB}(P/Q) = \mathrm{PROB}(P/Q\&R)$. Note that this immediately implies an analogous principle of irrelevance for causal probability:

(CIR) For any $P,Q,R$, it is defeasibly reasonable to expect that $\text{C-PROB}_A(P/Q\&R) = \text{C-PROB}_A(P/Q)$.

Now let us apply these observations to the computation of causal probabilities. (IR) gives us a defeasible presumption that actions are statistically irrelevant to their backgrounds, in which case it follows by theorem 2 that $\text{C-PROB}_A(O) = \mathrm{PROB}(O/A)$. In other words, it is then defeasibly reasonable to expect classical decision theory to yield the correct prescriptions. Causal decision theory only yields different prescriptions in the unusual case in which actions are statistically relevant to their backgrounds. These are cases like the smoking gene case in which the action and the possible outcome have common causal ancestors. As has been noted repeatedly in the literature, these cases are unusual.

In a case in which an action is known to be statistically relevant to some elements of its backgrounds, it follows from (IR) that it is defeasibly reasonable to expect that all other elements of its backgrounds are statistically independent of the action. Then it follows by theorem 2 that we can confine our attention to just those elements of the backgrounds to which the action is known to be statistically relevant, and hence deal with very small backgrounds. If more statistical relevance is found later, then the computation must be revised, but it is always defeasibly reasonable to expect that such a recomputation will not be necessary.

For some purposes, the preceding remarks put the cart before the horse. They suggest that $\text{C-PROB}_A(O)$ will be computed by first computing $\text{PROB}(O/A)$. In fact, the converse is likely to be true. The characterization of $\text{C-PROB}_A(O)$ in terms of scenarios provides what is in effect a recursive characterization, enabling us to compute causal probabilities by propagating them through time. If the case is classical, this is simultaneously a computation of $\text{PROB}(O/A)$. If the case is not classical, actions affect the probabilities of events occurring earlier than themselves. That does not make the causal probabilities harder to compute, but it can make $\text{PROB}(O/A)$ much harder to compute.

## 10. Conclusions

Some form of causal decision theory is required to handle the counterexamples to classical decision theory growing out of the Newcomb problem. As Lewis observes, the different causal decision theories that have been proposed are closely related to one another. In particular, they define causal probability by reference to concepts like causal dependence. Causal dependence is a philosophically problematic concept. It may be possible to define it in a way that makes it unproblematic, but at this stage its analysis is controversial, as are its logical properties. Accordingly, it seems undesirable to use it as a primitive constituent of an analysis of causal probability. This chapter has proposed that we can replace the appeal to causal dependence by appeal to temporal relations and statistical relevance between mixed physical/epistemic probabilities. The basic idea is simply that causes propagate through the world in temporal order. The resulting analysis handles the known counterexamples to classical decision theory in essentially the same way Skyrms' theory does, but without appealing to vaguely understood concepts like causal dependence.

# Part III

# Decisions

# 9
# Rational Choice and Action Omnipotence

## 1. Actions and the Optimality Prescription

We are systematically working our way through a series of objections to the optimality prescription. Thus far we have considered the familiar problems that gave rise to causal decision theory, and I have proposed that they can be handled by defining expected values in terms of causal probabilities, where the latter are as defined in chapter 8. In this chapter, I turn to a less familiar objection. This is that the optimality prescription ignores the fact that we may not know for certain, at the time we make a decision, whether we will be able to successfully perform the action we choose. It will be urged that in order to handle this we must further refine the definition of expected value.

Philosophers who are not decision theorists often assume uncritically that the "actions" of the optimality prescription are ordinary actions. In this they are certainly encouraged by the authors of the classical works. If we look at those works, there is no textual reason to doubt that these were the kinds of actions they concerned. For example, Jeffrey (1965) uses examples of choosing whether to *buy a weekend admission ticket to the beach* or *pay admission daily*, whether to *arm with nuclear weapons* or *disarm*, and whether to *take a bottle of red wine to a dinner party* or *take a bottle of white wine*. Savage (1954) uses the example of making an omelet and deciding whether *to break a sixth egg into a bowl already containing five good eggs*, or *to break the sixth egg into a saucer and inspect it before adding it to the bowl of eggs*, or *to discard the sixth egg without inspection*.

On reading the classical works, it certainly seems that the optimality prescription is to be applied to ordinary actions. But I have been surprised to discover (in conversation) that some contemporary decision theorists do not see things that way. They regard classical decision theory as an abstract mathematical theory, characterized by its axioms, and in principle immune from philosophical objections regarding its prescriptions. On this construal, to get any concrete prescriptions out of classical decision theory, we need a second theory that tells us how the first theory is to be interpreted. In particular, we need a theory telling us what the "decision-theoretic actions" are. On this view, classical decision theory just tells us that *something* is to be evaluated in terms of its expected value, and leaves open how that is to be used by a theory of rational choice. If we supplement classical decision theory with a theory about what decision-theoretic actions are, we get a theory of rational choice, but if philosophical objections are raised to the prescriptions of that theory of rational choice, they do not bear on decision theory—only on the theory of how it is to be interpreted.

I have no objection if this is the way one wants to understand classical decision theory. However, understood in this way, classical decision theory does not constitute a theory of rational choice. To get a theory of rational choice, we must conjoin classical decision theory with a second theory about what decision-theoretic actions are. This chapter makes three points. The first is that we cannot solve the problem of rational choice by identifying decision-theoretic actions with ordinary actions. If we do, the optimality prescription makes intuitively unreasonable prescriptions. The second point is that we can avoid the intuitive counterexamples by replacing the optimality prescription with a prescription that evaluates ordinary actions in terms of a more complex measure than expected value. The third point is that the resulting theory of rational choice is equivalent to applying the optimality prescription to something more complex than ordinary actions—what I call "conditional policies". So if we adopt the abstract understanding of classical decision theory, we can regard the conclusion of this chapter as proposing an alternative interpretation of classical decision theory wherein decision-theoretic actions are not actions at all, but rather conditional policies.

## 2. Action Omnipotence

Our target is a theory of rational choice between ordinary actions like *going to a movie* or *staying home and reading a book*. If the optimality prescription is to provide a direct answer to the question of how to make such choices, we must take the decision-theoretic actions to which the optimality prescription applies to be ordinary actions. However, if we understand the optimality prescription in this way, as being concerned with choices between ordinary actions, it is subject to a simple difficulty. The problem is that the optimality prescription would only be reasonable for actions that can be performed infallibly. This is the assumption of *action omnipotence*. To see that this is indeed required to make the optimality prescription defensible, consider a simple counterexample based on the failure of the assumption. Suppose that I am choosing between going to a movie or staying home and reading a book. I may decide that I would get more pleasure out of going to the movie, and so if that is the only source of value relevant to the decision, classical decision theory prescribes going to the movie. But now suppose my only way of going to the movie is to take a bus, and I know that there is talk of a bus strike. Suppose I believe that there is a 50% chance that there is a bus strike now in effect, and so there is only a 50% chance that I will be able to go to the movie. This is surely relevant to my decision whether to go to the movie, but it does not affect the expected value of *going to the movie*. That expected value is a weighted average of the values of the possible outcomes that will result if I actually do go, and takes no account of the possibility that I will be unable to go.

When faced with a problem like this, practical decision analysts simply reformulate the problem. The strategy is to take the uncertainty out of the action and move it into the consequences by taking the decision problem to be a choice between something like *trying to go to the movie* and *staying home and reading a book*. One might suppose that this should be regarded as a built-in constraint on classical decision theory—decision-theoretic actions must be infallibly performable. I have no objection if one wishes to stipulate this, although as remarked above it is hard to justify this on textual grounds. The examples used by the authors of the classical texts do not involve infallibly performable actions. They involve actions like *taking a bottle of red wine to a dinner party*.

Suppose we do stipulate that the optimality prescription is only to be applied to infallibly performable actions. That avoids the counterexample, but remember that our topic is a theory of rational choice, and this concerns ordinary actions—not just infallibly performable actions. Ordinary actions (of the sort Jeffrey and Savage used as examples) are not guaranteed to be infallibly performable. There are basically two ways a theory of rational choice might deal with the failure of action omnipotence. We can insulate the optimality prescription from the difficulties stemming from the failure of action omnipotence by restricting it to infallibly performable actions, but only at the cost of making it no longer provide a direct answer to questions of rational choice between ordinary actions. At the very least, it then needs to be supplemented with an account of how choices between ordinary actions are to be made by reference to choices between infallibly performable actions. Alternatively, we can look for a "decision-theory-like" theory of rational choice that applies directly to ordinary actions but modifies the way in which actions are evaluated. I will consider the former strategy first, but argue that it fails because there are is no guarantee that there are any infallibly performable actions. I will take that to motivate the second strategy.

## 3. Restricting the Scope of the Optimality Prescription

The first suggestion is that although the optimality prescription makes incorrect prescriptions if we take it to apply to actions that are not infallibly performable, we can avoid this difficulty by restricting the range of actions to which the prescription is applied. I will argue that this simple strategy fails. Once we make the notion of infallible performability more precise, it will turn out that there are no kinds of actions such that all actions of those kinds are guaranteed to be infallibly performable. Furthermore, in real cases there may be no actions that are, even as a matter of fact, infallibly performable, so if the optimality prescription is only applicable to infallibly performable actions, there will be cases in which it gives us no help in our decision making. To defend these claims we must survey the possible candidates for infallibly performable actions. There are few concrete proposals in the literature regarding what actions are infallibly performable, so we are largely on our own in looking for candidates. I will consider four ways of trying to restrict the scope of the optimality prescription so as to avoid the problem of action omnipotence.

### *3.1 Acts and Actions*

To sort out the problems for applying classical decision theory to rational choice, we need a clearer understanding of actions. Let us begin with a type/token distinction. I will say that *acts* are individual spatio-temporally located performances. (I mean to include mental acts here.) I will take *actions* to be act-types. In rational decision making it is actions we are deliberating about. That is, we are deciding what type of act to perform.

The individuation of acts is philosophically problematic. If I type the letter "t" by moving my left index finger, are the acts of typing the letter "t" and moving my left index finger two acts or one? I do not think that this question has a predetermined answer. We need some legislation here. On the one hand, acts might be broadly individuated in terms of what physical or mental "movements" the agent makes. In this sense, I only performed one act.[1] However, some philosophers have insisted that two acts are performed in this case. One way to achieve this result is to take the type that is part of the specification of the act to be determined by the agent's intentions.[2] On this construal, if I have the intention to type the letter "t" and I also have the intention to move my left index finger, then there are two acts. On the other hand, if I type the letter "t" inadvertently when I move my finger, that is not an act that I performed. Acts individuated partly by reference to the agent's intentions are *acts narrowly individuated.* For present purposes, it is more convenient to take acts to be narrowly individuated, so that will be the convention adopted in this book. If I want to talk about acts individuated broadly, I will say that explicitly. It is important to realize that this really is just a convention—not a substantive philosophical claim. If one takes broadly individuated acts to be basic, narrowly individuated acts can be identified with ordered pairs ⟨*act*,*action*⟩ where *act* is a broadly individuated act of type *action*, where the type is narrowly specified.

### *3.2 Levels of Actions*

Actions exhibit a continuous range of abstractness. Consider the actions *wiggle your finger, walk across the room, make a cup of coffee, vacation in Costa Rica, save the world.* I often perform a high-level act *by* performing a lower-level act or a sequence of lower-level acts. Goldman (1970) called this "level generation". For instance, I make a cup of coffee by performing the sequence

---

[1] This view is endorsed by Anscombe (1958), Davidson (1963), Schwayder (1965), and others.

[2] Danto (1963) and Goldman (1970) endorse the "two acts" theory, although their interest is in more than intentional acts, so they cannot include an intention in the act specification. My interest here is exclusively in intentional acts, because they alone are products of rational deliberation.

of acts consisting of walking into the kitchen, putting water and coffee in the coffee pot, turning it on, waiting a while, and then pouring the brewed coffee into a cup. In turn, I put water in the coffee pot by picking it up, putting it under the tap, turning on the water, waiting until the water reaches the appropriate level in the pot, turning off the tap, and setting the coffee pot down on the counter. We can progress to lower and lower levels of acts in this way, but eventually we will reach acts like *grasp the handle of the coffee pot* or *raise my arm* that I can perform "directly"—without performing them by performing some simpler act.

As I am construing it, level generation is a relationship between acts, not actions. The literature on action theory defines *basic acts* to be acts that are not performed by performing another act. For example, wiggling my finger will normally be a basic act. Note that the basic/non-basic distinction only makes sense for acts narrowly individuated—not for acts broadly individuated. If I type the letter "t" by moving my finger, there is only one broadly individuated act performed. Note also that the distinction between basic and nonbasic acts makes equally good sense when applied to artificial agents like robots. There are some acts they can perform directly—moving their effectors—and others they can only perform by performing some simpler acts.

A nonbasic act can be performed by performing another nonbasic act. For example, you can turn on the light by throwing a switch. But you throw the switch by moving your finger, which is a basic act. So it follows that you also turn on the light by moving your finger (level-generation is transitive). On the assumption that there cannot be an infinite chain of level-generation, it follows that a nonbasic act is always performed by performing some basic act or sequence of basic acts.

We can think of actions (act types) as characterized by the range of sequences of basic acts by which acts of that type can be performed. This will generally be an open-ended range. For example, I can turn on the light by throwing a switch, but with sufficient ingenuity I can always think of new ways of turning on the light, for example, Rube Goldberg devices, training my dog to rub against the switch, wiring in new power sources, and so on.

It is important to keep in mind that the basic/non-basic distinction is a distinction between acts, not actions. We might try defining a *basic action* to be an action (an act type) for which every act of that type is a basic act. However, so defined, there are no physical basic actions. For example, *wiggling my finger* is normally a basic act. But if my finger is paralyzed, I might wiggle it with my other hand. In that case, I wiggle my finger by doing something else, so it is not a basic act. Similarly, if an undersea robot has a malfunctioning arm, it might move that arm with its other arm in order to use the grasper on the broken arm. A more useful notion is that of a *potentially basic action,* which is an act type that *can* have tokens that are basic acts. Given the narrow individuation of acts, most actions are not potentially basic actions. At least for humans, *fixing a cup of coffee* is not, but *wiggling my finger* is.

Most mental actions are not basic either. For example, multiplying 356 by 123 is something I do by performing a sequence of simpler multiplications. Perhaps I usually multiply 3 and 5 by performing a basic act, but I need not. I might perform it as sequential addition. So even that is only a potentially basic action. However, there are a few mental actions that plausibly cannot be performed by doing something else. For example, *recalling my mother's maiden name* might be a basic action. I can do things to cause me to recall my mother's maiden name, but I do not recall it *by* doing those things, at least not in the sense that recalling it is constituted by doing those other things. So there may be some basic mental actions.

### 3.3 Basic Actions and Action Omnipotence

The problem we have noted for classical decision theory seems initially to pertain primarily to high-level actions. They can be difficult to perform, or even impossible to perform in the present circumstances, and that ought to be relevant to their decision-theoretic evaluation. It is tempting to try to avoid this difficulty by restricting the dictates of decision theory to low-level actions. Even if this were to work, it would not provide a fully adequate theory of rational choice, because the practical decisions we make generally involve choices between fairly high-level actions. For example, we want to capture reasoning like the following:

> Should I make a cup of coffee, or work in the garden? I would get more immediate gratification out of having a cup of coffee, but it would make me edgy later. If I work in the garden, I will sleep well tonight. So perhaps I should do the latter.

So we cannot save classical decision theory as a theory of rational choice by insisting that it only makes recommendations about low-level actions. Decisions about actions like making a cup of coffee are the kinds of decisions a theory of rational choice is supposed to be about. If the optimality prescription is inapplicable to these choices, then it does not, at least directly, constitute a theory of rational choice. However, it might be suggested that a theory of rational choice for high-level actions could be based indirectly on a theory of rational choice for low-level actions, and classical decision theory might provide the latter.

I have heard it suggested (not in print, but then, little about this topic has been suggested in print) that action omnipotence holds for basic actions, and so classical decision theory should be restricted to those. But as we have seen, there are no basic physical actions. And actions that are merely potentially basic are not infallibly performable. If your finger has been injected with a muscle paralyzer, you may not be able to wiggle your finger. (You might be able to wiggle it by doing something else, for example, moving it with your other hand, but if you are sufficiently constrained you may be unable to move it at all.) This can be relevant to practical decisions. Suppose I offer you the following choice. I will give you $10 if you wiggle your left index finger in one minute, but I will give you $100 if instead you wiggle your right index finger in ten minutes. Decisions to act in a certain way must always be made in advance of acting, but the one minute/ten minute difference in this example has the consequence that if you decide to wiggle your right index finger rather than the left, you must do so at least nine minutes before you perform the action. The hitch is that your right index finger is currently paralyzed and you are unsure whether the paralysis will have worn off by the time you are to wiggle it. Your assessment of how likely you are to be able to wiggle your right index finger at the appropriate time is surely relevant to your rational decision, but restricting classical decision theory to potentially basic actions yields a theory that makes no provision for this. It dictates instead that you should choose to wiggle your right index finger in ten minutes, even if it is improbable that you will be able to do that.

### 3.4 Deciding

Henry Kyburg recently suggested in conversation that the problem of action omnipotence may be handled by taking the actions to be evaluated decision-theoretically to be *decidings*. I suggested something similar in my (1995). However, it takes little reflection to realize that we are sometimes unable to decide. We have all experienced cases of indecisiveness, especially when the decision is very important. Think about how much difficulty people have in deciding whether to get married, or to take a new job. Sometimes indecisiveness is perfectly rational. For example, there may be a crucial piece of information that the agent lacks but expects to get later. Suppose you are to pick up a friend at the airport, but you do not know which of two airports he will be flying into. He has promised to e-mail his flight information to you later in the day. In this case, you cannot decide now to which airport you will drive.

### 3.5 Willing

As noted in chapter 1, philosophers sometimes claim that actions are initiated by engaging in a mental action they call "willing". Perhaps choices should be between willings.[3] However, I am not sure what willings are supposed to be. We do talk about willing ourselves to do something, but I doubt this can carry much theoretical weight. Do we have to will ourselves to will? If so, we seem to be threatened with an infinite regress. On the other hand, if we can initiate a willing without willing to will, why can't we initiate a finger wiggling without willing that?

Whatever willings amount to, if we can appeal to our ordinary talk of willing ourselves to do things, then it is clear that we are sometimes unable to will ourselves to act in a certain way. Consider a person who is afraid of needles but is trying to give himself a shot. He may be unable to will himself to stick the needle in his arm. A more dramatic example is provided

---

[3] Joyce (1998, pg. 57) suggests that choices should be between "pure, present exercises of the will", and he takes *resolvings* to be of this nature. This might be construed as a vesion of the "decidings" account just discussed, or as a version of the "willings" account.

by a recent newspaper report of a hiker in southern Utah. He was hiking alone in a canyon, and a boulder moved unexpectedly and pinned his arm. After three days he was out of water, and knew he would die if he did not extricate himself. He saved himself by amputating his arm with a pen knife. What makes this story remarkable is that most people think to themselves, "I couldn't do that." They could not will themselves to cut off their own arm in such a situation.

### *3.6 Trying*

At one time I thought that although we cannot always perform an action, we can always try, and so classical decision theory should be restricted to tryings. This is also suggested by a passing remark in Jeffrey (1985, p. 83). Rather than choosing between moving my left index finger and moving my right index finger, perhaps my choice should be viewed as one between trying to move my left index finger and trying to move my right index finger. That handles the example of the paralyzed muscle nicely. Assuming that the probability is 1 that I can (under the present circumstances) move my left index finger if I try, then the expected value of trying to do it is $10. The expected value of trying to move my right index finger is $100 times the probability that I will be able to move it if I try. If that probability is greater than .1, then that is what I should choose to do.

At first glance, it seems that trying might be a basic action and infallibly performable. Then we could salvage the optimality prescription by saying that talk of choices between actions is really short for talk of choices between trying to perform actions. Or we could say that we are choosing between actions, but actions are to be evaluated in terms of the expected values of trying to perform them.

However, things are not so simple. First, trying to perform an action is at least not usually a basic action. It is something one can do by doing something else—normally something physical. For example, I may try to move a boulder by placing a pry bar under it and leaning on the pry bar. Of course, if action omnipotence held for trying, it would make no difference whether trying is a basic action. In that case we could base decision theory upon comparisons of the expected values of tryings. Unfortunately, it is not true that we can always try. Suppose I show you a wooden block, then throw it in an incinerator where it is consumed, and then I ask you to paint it red. You not only cannot paint it red—you cannot even try to do so. There is nothing you could do that would count as trying.

The state of the world makes it impossible for you to paint the block red. But what makes it impossible for you to try to paint it red is not the state of the world but rather your beliefs about the state of the world. For example, suppose I fooled you and you just *think* I threw the block in the incinerator. Then although the action of painting the block red is one that someone else could perform, *you* cannot even try to do it.[4] Conversely, if I did destroy the

[4] Note that trying is intensional. You might try to paint the block on the table red without

block but you do not believe that I did, you would not be able to paint it but you might be able to try. For instance, if you believe (incorrectly) that the block is at the focus of a set of paint sprayers activated by a switch, you could try to paint the block by throwing the switch. These examples illustrate that what you can try to do is affected by your beliefs, not just by the state of the world.

What exactly is it to try to perform an action? If *A* is an action that is not potentially basic, you try to perform *A* by trying to execute a plan for doing *A*. This plan may either be the result of explicit planning on your part, or the instantiation of a plan schema stored in memory (for example, how to get home from the office), or the instantiation of procedural knowledge that is "compiled in" (for example, how to ride a bicycle). One way in which you may be unable to try to do *A* is that, at the time you are supposed to act, you have no plan for doing *A*. (You need not have settled on a plan at the time you decide to do *A*. For example, I may decide to fly to LA before deciding what flight to take.) Another way in which it can happen that you are unable to try is that your plan for *A*-ing requires certain resources, and when the time comes for executing it you do not have (or, perhaps, believe you do not have) the resources. For instance, I may not be able to try to paint the block because I do not have any paint. Finally, plans often contain epistemic contingencies. At the time you make the plan you may expect to have the requisite knowledge when it is time to execute the plan, but if you do not then you may not even be able to try to *A*. For instance, you may be unable to try to paint the block because you do not know which one it is (in a pile of blocks), or you do not know where it is, or you do not know where the paint is.

For a potentially basic action *A*, you can either try to perform *A* by trying to execute a plan for doing *A*, or by trying to perform *A* in a "basic way". It is unclear to me whether you can always try to perform a potentially basic action in a "basic way". For example, if you know that your finger has been amputated, can you still try to wiggle it? This might be handled by saying that wiggling your finger is no longer a potentially basic action. But if that is right, rational choice seems to require you to have beliefs about whether your future actions are going to be potentially basic at the time you try to perform them. For instance, your finger may be intact now but you might have to worry about its being amputated before the time you plan to wiggle it.

It is important to remember that decisions to perform actions are always made in advance of the time the action is to be performed. It makes no literal sense to talk about deciding whether to do something *now*. If "now" literally means "at this very instant", then you are either already performing the action or refraining from performing it. It is too late to make a decision. In rational choice, there must always be some time lag between the decision and the action. The time lag can be a matter of years, but even when it is a

---

realizing it is the block you thought destroyed, but then you have not tried to paint the block you thought destroyed.

matter of seconds this has the consequence that you cannot be absolutely certain either that you will be able to perform the action or that you will be able to try to perform the action. At least often, your inability to perform an action may result from your epistemic state, but it is not your current epistemic state that determines whether you will be able to try—it is your epistemic state at the time of performing the action. So even if one were to suppose that we can be absolutely certain about our current epistemic state, it would not follow that we can be certain about whether we will be able to perform the action at some future time. A consequence of this is that *trying-to-A* is not infallibly performable, and so does not provide a candidate for the kinds of infallibly performable actions required to make the prescriptions of classical decision theory reasonable.

In response to these observations I sometimes get the protest, "That is not what I mean by 'trying'—I mean something that is infallibly performable." But the philosopher who simply asserts this and stops without further elaborating his view is just refusing to come to grips with the problem. If he thinks there is something he *can* mean by "trying" that is infallibly performable, he had better tell us what it is.

### 3.7 *Giving Up on Infallibly Performable Actions*

Action omnipotence fails in two ways—we may fail to perform an action when we try, and we may not even be able to try. This suggests that we might define infallible performability more precisely in terms of probabilities. What it requires is that (1) the probability of successfully performing an act of type *A* if we try is 1, and (2) the probability of being able to try is also 1. More simply, $\textbf{PROB}(A/try\text{-}A) = 1$ and $\textbf{PROB}(can\text{-}try\text{-}A) = 1$. These probabilities will vary from circumstances to circumstances. If, as I think we should, we take the probabilities involved to be mixed physical/epistemic probabilities, then there will be circumstances in which they are both 1. This follows from the fact that if one knows that *P* is true, then $\textbf{PROB}(P) = 1$. There will certainly be circumstances in which I know both that I can try to do something and that if I try I will succeed. For instance, right now I know that I can try to type the letter "t" on my keyboard, and I also know that if I try I will succeed. So this action is currently infallibly performable for me. Of course, we can imagine other circumstances in which it is not.

My claim is not that there are never any infallibly performable actions (types of acts), but only that this depends on the circumstances, and there are no actions that are necessarily infallibly performable under all circumstances. For every type of act, we can imagine circumstances in which it is not infallibly performable. It might be supposed that problems arising from the failure of action omnipotence can always be avoided by framing our decision problems correctly—in terms of infallibly performable actions. However, there is no reason to expect that this will always be possible. For instance, suppose I am considering flying to Copenhagen on Tuesday. At this late date there is no guarantee that I will be able to get a seat. I might be able to try to fly to Copenhagen by putting myself on the standby list. But I may not even be able to try. Last-minute bookings can be expensive. I may not be able to afford a ticket. We would naturally express this by saying that I cannot buy a ticket. I cannot even try to buy a ticket.

The upshot is that sometimes ordinary actions will be infallibly performable, but often they will not. And when they are not, we may still have to decide whether to perform them. We might express this by saying that we have to decide whether to *try* to perform them, but all that is doing is giving lip service to the failure of action omnipotence. Trying to perform the action may not be infallibly performable either. If we look hard enough we may be able to find some infallibly performable actions that bear some relationship to what we want to do, but even if we can it is unclear how to use that fact in solving our decision problem. And there appears to be no logical reason why there should be any relevant acts that are infallibly performable. A perfectly reasonable decision problem could be fraught with uncertainty all the way down.

My conclusion is that we cannot solve the problem of constructing a theory of rational choice by merely insisting that the optimality prescription be restricted to infallibly performable actions. There may be no infallibly performable actions related to our decision problem, and even if there are, deciding whether to perform *those actions* may fall far short of solving the original decision problem. The lesson to be learned is that the performability of actions comes in degrees, as the probabilities PROB(*A* / *try-A*) and PROB(*can-try-A*) vary. We must seek a different principle of rational choice that applies to ordinary actions but takes account of the fact that these probabilities can be less than 1. Perhaps this can be done by changing the measure that is to be optimized in assessing actions. The next section explores this possibility.

## 4. Expected Utility

How might we change the way in which we assess actions so that it accommodates the failure of action omnipotence? It is tempting to suppose that we should simply discount the expected value of performing an action by the probability that we will be able to perform it if we try:

$$\mathbf{EV}(A)\cdot\mathrm{PROB}(A/\,try\text{-}A).$$

That does not work, however. The values we must take account of in assessing an action include (1) the values of any goals achieved by performing the action, (2) execution costs, and (3) side-effects that are not among the goals or normal execution costs but are fortuitous consequences of performing or trying to perform the action under the present circumstances. The goals will presumably be consequences of successfully performing the action, but execution costs and side effects can result either from successfully performing the action or from just trying to perform it. For example, if I try unsuccessfully to move a boulder with a pry bar, I may expend a great deal of energy, and I might even pull a muscle in my back. These are execution costs and side effects, but they attach to the trying—not just to the successful doing. These costs are incurred even if one tries to $A$ but fails, so their contribution to the assessment of the action should not be discounted by the probability that the action will be performed successfully if it is attempted.

To include all of these values and costs in our assessment of the action, we might look at the expected value of trying to perform the action rather than the expected value of actually performing it:

$$\mathbf{EV}(\textit{try-A}).$$

This will have the automatic effect of discounting costs and values attached to successfully performing the action by the probability that we will perform it if we try, but it also factors in costs and values associated directly with trying. However, more is relevant to the assessment of an action than the expected value of trying to perform it. As we have seen, we may not be able to try to perform an action. It seems apparent that in comparing two actions, if we know that we cannot try to perform one of them, then it should not be a contender in the choice. That might be handled by excluding it from the set **A** of alternatives. But more generally, we may be uncertain whether we will be able to try to perform the action at the appropriate time. For example, consider a battle plan in which we are considering using an airplane to attack the enemy. However, the airplane is parked on an undefended airstrip and there is the possibility that the enemy will destroy the airplane before we can use it. If the plane is destroyed, we will be unable to try to attack the enemy by using it. Clearly, the probability of the plane's being destroyed should affect our assessment of the action when we compare it with alternative ways of attacking the enemy.

The obvious suggestion is that we should discount the expected value of trying to perform an action by the probability that, under the present circumstances, we can try to perform it:

$$\mathbf{EV}(\textit{try-A})\cdot\mathbf{PROB}(\textit{can-try-A}).$$

A surprising qualification is required, however. Rather than looking at the expected value of trying to perform an action $A$, I will argue that we must consider the conditional expected value of trying to perform $A$ *given that* the agent can try to perform $A$. Conditional expected values were defined as follows:

$$\mathbf{EV}(A/P) = \sum_{O\in\mathbf{O}} \mathrm{U}(O)\cdot\textbf{C-PROB}_A(O/P).$$

This is the expected value of the action given the assumption that $P$ is true. In non-causal decision theory, conditionalizing on the agent being able to try to perform $A$ makes no difference, because the conditional expected value of trying to perform $A$ given that the agent can try to perform $A$ is defined in terms of **PROB**($O$/ *the agent tries to perform A & the agent can try to perform A*). If the agent tries to perform $A$, it follows logically that he can try to perform $A$, so this probability is the same as **PROB**($O$/ *the agent tries to* *perform A*), and hence the conditional expected value is the same as the unconditional expected value. However, in causal decision theory, this equivalence fails. By definition, where $\mathbf{B}$ is the set of backgrounds for $A$:

$$\textsf{C-PROB}_{try\text{-}A}(O/\textit{the agent can try to perform } A)$$
$$= \sum_{B \in \mathbf{B}} [\textsf{PROB}(B\ /\textit{the agent can try to perform } A)$$
$$\cdot\textsf{PROB}(O\ /B\ \&\ \textit{the agent tries to perform } A\ \&\ \textit{the agent can try to perform } A)].$$

As before,

$$\textsf{PROB}(O\ /B\ \&\ \textit{the agent tries to perform } A\ \&\ \textit{the agent can try to perform } A)$$
$$= \textsf{PROB}(O/B\ \&\ \textit{the agent tries to perform } A)$$

but it need not be the case that $\textsf{PROB}(B\ /\textit{the agent can try to perform } A) = \textsf{PROB}(B)$. Knowing that the agent can try to perform $A$ may alter the probability of relevant earlier states of affairs. To illustrate, think again of the battle plan. If the enemy destroys the airplane before we can use it, we will not even be able to try to use it. Now consider the causal probabilities in this example. If the plane has not been destroyed by the time we plan to use it (a necessary condition for our being able to try to attack the enemy with it), that may increase the probability that the enemy's air force has been crippled by earlier attacks, and the probability of the new attack being successful will thereby be increased because the enemy will be less likely to be able to repel it. That is,

$$\textsf{PROB}(\textit{the enemy air force was crippled earlier}\ /\textit{we can try to attack with the airplane})$$
$$> \textsf{PROB}(\textit{the enemy air force was crippled earlier})$$

with the result that

$$\textsf{C-PROB}_{try\ to\ attack}(\textit{attack will be successful}/\textit{we can try to attack})$$
$$> \textsf{C-PROB}_{try\ to\ attack}(\textit{attack will be successful}).$$

This raises the expected value of trying to attack with the airplane given that one can try (that is, given that the airplane is still there). Given this difference, it seems clear that it is the conditional causal probability that should be employed in evaluating the attempt to attack. This suggests that action should be evaluated in terms of the conditional expected value of trying to perform it given that one can try to perform it:

**EV**(*try-A*/*can-try-A*)·**PROB**(*can-try-A*).

This still does not quite work, however. Suppose you are in a situation in which you get at least $10 no matter what you do. You get another $10 if you do *A*, but you have only a 50% chance of being able to try to do *A*. If you can try to do *A*, you have a 100% chance of succeeding if you try. If, instead of doing *A*, you do *B*, you will get $1 in addition to the $10 you get no matter what. Suppose you are guaranteed of being able to try to do *B*, and of doing *B* if you try. Given a choice between *A* and *B*, surely you should choose *A*. You have a 50% chance of being able to try to do *A*, and if you do try you will get ten extra dollars. You have a 100% chance of being able to try to do *B*, but if you try to do *B* you will only get one extra dollar.

However, the only possible outcome of trying to do *A* is worth $20, and the only possible outcome of trying to do *B* is worth $11. So these are the conditional expected values of trying to perform *A* and *B*. If we discount each by the probability of being able to try to perform the action, the value for *A* is $10 and that for *B* is $11. This yields the wrong comparison. It is obvious what has gone wrong. We should not be comparing the total amounts we will get if we perform the actions, and discounting those by the probabilities of being able to try to perform the actions. Rather, we should be comparing the *extra amounts* we will get, over and above the $10 we will get no matter what, and discounting those extra amounts by the probabilities of being able to perform the actions. This is diagrammed in figure 9.1.

Apparently, in choosing between alternative actions, we must look at the expected values that will accrue specifically as a result of performing each action. This is not the same thing as the expected value of the outcome of the action, because some of the value of the outcome may be there regardless of what action is performed. The expected value that will accrue specifically as a result of trying to perform *A* is the difference between the conditional expected value of trying to perform *A* and the conditional expected value of not trying to perform any of the alternative actions (*nil*—the "null-action"). This is the *conditional marginal expected value* of trying to perform the action. It is this conditional marginal expected value that should be discounted by the probability of being able to try to perform the action.

A word about *nil*. At this point I am trying to remain neutral about as many aspects of classical decision theory of possible. In particular, I do not want to make any unnecessary assumptions about the set of alternative actions **A**. Some authors assume that **A** is an exhaustive list of alternatives, in the sense that, necessarily, the agent will perform one of the actions in **A**. One alternative is always to do nothing, so on this assumption **A** will contain the null action as one of its members. However, for reasons discussed in the next chapter, I think it is best not to assume that **A** is an exhaustive list of alternatives—it is just the set of alternatives the agent is considering. Thus I do not assume that *nil* will always be a member of **A**. That is not required for the way *nil* is used in computing marginal expected values. There it is just used as a computational device—not as one of the alternatives in **A**. If *nil* is not included in **A** then it is defined as not performing any member of **A**.

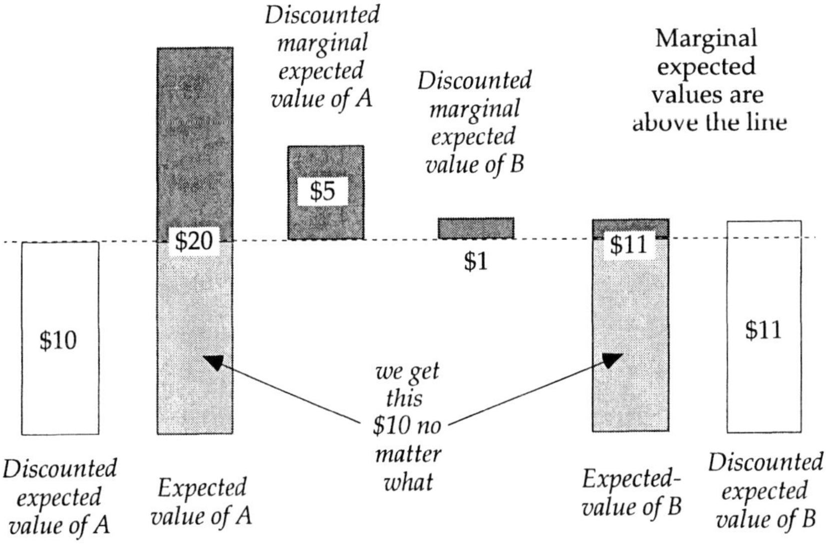


Figure 9.1. Comparing discounted expected values and discounted marginal expected values

Putting this all together, I propose that we define the *expected utility* of an action to be the conditional marginal expected value of trying to perform that action, discounted by the probability that we can try to do that:

$$\mathbf{EU}(A) = \textsf{PROB}(\textit{can-try-A})\cdot[\mathbf{EV}(\textit{try-A}\,/\,\textit{can-try-A}) - \mathbf{EV}(\textit{nil}\,/\,\textit{can-try-A})].$$

If the probability that the agent can try to perform *A* is 0, the conditional expected value of trying to perform *A* is undefined, but in that case let us just stipulate that **EU**(*A*) = 0. In classical decision theory, the terms "the expected value of an action" and "the expected utility of an action" are often used interchangeably, but I am now making a distinction between them. My proposal is that we modify the optimality prescription by taking it to dictate choosing between alternative actions on the basis of their expected utilities. With this change, the optimality prescription is able to handle all of the above examples in a reasonable way, without restricting it to a special class of particularly well-behaved actions.

However, this definition of the expected utility of an action seems a bit *ad hoc*. It was propounded to yield the right answer in decision problems, but why is this the right concept to use in evaluating actions? It will be shown in the next section that it has a simple intuitive significance. Comparing actions in terms of their expected utilities is equivalent to comparing the expected values of conditional policies of the form *try to do A if you can try to do A*.

## 5. Conditional Policies and Expected Utilities

Decision theory has usually focussed on choosing between alternative actions available to us *here and now*. A generalization of this problem is important in some contexts. We sometimes make *conditional decisions* about what to do if some condition $P$ turns out to be true. For instance, I might deliberate about what route to take to my destination if I encounter road construction on my normal route. Where $P$ predates $A$, *doing A if P* (and performing none of the alternative actions otherwise) is a *conditional policy*.[5] Conditional decisions are choices between conditional policies. Presumably we should make such conditional decisions on the basis of the expected values of the conditional policies. We can define

$$\mathbf{EV}(A \textit{ if } P) = \sum_{O\in\mathbf{O}} \mathrm{U}(O)\cdot\textbf{C-PROB}_{A \textit{ if } P}(O).$$

To complete this definition, we must define the causal probability of $O$ conditional on execution of the conditional policy. The following should be true:

$$\textbf{C-PROB}_{A \textit{ if } P}(O)$$

$$= \textbf{C-PROB}_{A \textit{ if } P}(P)\cdot\textbf{C-PROB}_{A}(O/P) + \textbf{C-PROB}_{A \textit{ if } P}(\sim P)\cdot\textbf{C-PROB}_{nil}(O/\sim P).$$

This does not yet constitute a definition, because $\textbf{C-PROB}_{A \textit{ if } P}$ occurs on the right side of the equation. However, $P$ predates $A$, so we should have $\textbf{C-PROB}_{A \textit{ if } P}(P) = \textbf{PROB}(P)$ and $\textbf{C-PROB}_{A \textit{ if } P}(\sim P) = \textbf{PROB}(\sim P)$. This allows us to turn the preceding principle into a definition:

$$\textbf{C-PROB}_{A \textit{ if } P}(O) = \textbf{PROB}(P)\cdot\textbf{C-PROB}_{A}(O/P) + \textbf{PROB}(\sim P)\cdot\textbf{C-PROB}_{nil}(O/\sim P).$$

We can define the expected value of a conditional policy straightforwardly as follows:

$$\mathbf{EV}(A \textit{ if } P) = \sum_{O\in\mathbf{O}} \mathrm{U}(O)\cdot\textbf{C-PROB}_{A \textit{ if } P}(O).$$

It is important to distinguish between the expected value of a conditional policy and a conditional expected value. The latter was defined as follows:

---

[5] One can generalize conditional policies in various ways, e.g., looking at policies of the form "*Do A if P and do B if ~P*". These generalizations are discussed in chapter 10. However, we only need this simple form of conditional policy for the present discussion.

$$\mathbf{EV}(A/P) = \sum_{O \in \mathbf{O}} \mathrm{U}(O)\cdot\textsc{c-prob}_A(O/P).$$

This is the expected value of the action given the assumption that $P$ is true. **EV**($A$ *if* $P$), on the other hand, is the expected value of doing $A$ if $P$ and doing nothing otherwise. The expected value of a conditional policy is related to conditional expected values as follows:

**Theorem 1:** $\mathbf{EV}(A\ if\ P) = \textsc{prob}(P)\cdot\mathbf{EV}(A/P) + \textsc{prob}(\sim P)\cdot\mathbf{EV}(nil/\sim P)$.

Proof:

$$\begin{aligned}\mathbf{EV}(A\ if\ P) &= \sum_{O \in \mathbf{O}} \mathrm{U}(O)\cdot\mathbf{PROB}_{A\ if\ P}(O)\\ &= \sum_{O \in \mathbf{O}} \mathrm{U}(O)\cdot[\mathbf{PROB}(P)\cdot\mathbf{PROB}(O/P\&A)\\ &\qquad + \mathbf{PROB}(\sim P)\cdot\mathbf{PROB}(O/\sim P\&nil)]\\ &= \sum_{O \in \mathbf{O}} \mathrm{U}(O)\cdot\mathbf{PROB}(P)\cdot\mathbf{PROB}(O/P\&A)\\ &\quad + \sum_{O \in \mathbf{O}} \mathrm{U}(O)\cdot\mathbf{PROB}(\sim P)\cdot\mathbf{PROB}(O/\sim P\&nil)\\ &= \mathbf{PROB}(P)\cdot\sum_{O \in \mathbf{O}} \mathrm{U}(O)\cdot\mathbf{PROB}(O/P\&A)\\ &\quad + \mathbf{PROB}(\sim P)\cdot\sum_{O \in \mathbf{O}} \mathrm{U}(O)\cdot\mathbf{PROB}(O/\sim P\&nil)\\ &= \mathbf{PROB}(P)\cdot\mathbf{EV}(A/P) + \mathbf{PROB}(\sim P)\cdot\mathbf{EV}(nil/\sim P).\ \blacksquare\end{aligned}$$

Decision theory has normally been concerned with expected values. However, the expected value of an action is defined to be the expected value of "the world" when the action is performed.[6] This includes values that would have been achieved with or without the action. As we saw in section 4, it is often more useful to talk about the conditional marginal expected value, which is the difference between the conditional expected value of the action and the conditional expected value of doing nothing. The conditional marginal expected value of an action measures how much value the action can be expected to *add* to the world. The general definition is:

$$\mathbf{MEV}(A/P) = \mathbf{EV}(A/P) - \mathbf{EV}(nil/P).$$

We can define the marginal expected value of a conditional policy analogously:

---

[6] This is the "official" definition, e.g., in Savage (1954), Jeffrey (1965), etc. However, in classical decision theory values can be scaled linearly without affecting the comparison of expected values, so in actual practice people generally employ marginal expected values in place of expected values. For comparing expected utilities (defined as above), such scaling can change the comparisons, so we must use marginal expected utilities rather than expected utilities.

$$\mathbf{MEV}(A \textit{ if } P) = \mathbf{EV}(A \textit{ if } P) - \mathbf{EV}(\textit{nil if } P).$$

The conditional policy *nil if P* prescribes doing *nil* if *P* and *nil* if ~*P*, so it is equivalent to *nil* simpliciter. Thus, we could just as well have defined:

$$\mathbf{MEV}(A \textit{ if } P) = \mathbf{EV}(A \textit{ if } P) - \mathbf{EV}(\textit{nil}).$$

It follows that comparing conditional policies in terms of their marginal expected values is equivalent to comparing them in terms of their expected values:

**Theorem 2:** $\mathbf{MEV}(A \textit{ if } P) > \mathbf{MEV}(B \textit{ if } Q)$ iff $\mathbf{EV}(A \textit{ if } P) > \mathbf{EV}(B \textit{ if } Q)$.

There is a simple relationship between the marginal expected value of a conditional policy and the conditional marginal expected value:

**Theorem 3:** $\mathbf{MEV}(A \textit{ if } P) = \mathbf{PROB}(P)\cdot\mathbf{MEV}(A/P)$

Proof:

$\mathbf{MEV}(A \textit{ if } P)$

$= \mathbf{EV}(A \textit{ if } P) - \mathbf{EV}(\textit{nil if } P)$

$= \mathbf{PROB}(P)\cdot\mathbf{EV}(A/P) + \mathbf{PROB}(\sim P)\cdot\mathbf{EV}(\textit{nil}/\sim P)$
$\quad - \mathbf{PROB}(P)\cdot\mathbf{EV}(\textit{nil}/P) - \mathbf{PROB}(\sim P)\cdot\mathbf{EV}(\textit{nil}/\sim P)$

$= \mathbf{PROB}(P)\cdot\mathbf{EV}(A/P) - \mathbf{PROB}(P)\cdot\mathbf{EV}(\textit{nil}/P)$

$= \mathbf{PROB}(P)\cdot[\mathbf{EV}(A/P) - \mathbf{EV}(\textit{nil}/P)]$

$= \mathbf{PROB}(P)\cdot\mathbf{MEV}(A/P)$. ■

We defined $\mathbf{EU}(A) = \mathbf{PROB}(\textit{can-try-A})\cdot\mathbf{MEV}(\textit{try-A}/\textit{can-try-A})$, so by theorem 1:

**Theorem 4:** $\mathbf{EU}(A) = \mathbf{MEV}(\textit{try-A if can-try-A})$.

Hence by virtue of theorem 2:

**Theorem 5:** $\mathbf{EU}(A) > \mathbf{EU}(B)$ iff $\mathbf{EV}(\textit{try-A if can-try-A}) > \mathbf{EV}(\textit{try-B if can-try-B})$.

In other words, comparing actions in terms of their expected utilities is equivalent to comparing conditional policies of the form *try-A if can-try-A* in terms of their expected values. This, I take it, is the explanation for why examples led us to this definition of expected utility. Due to the failure of action omnipotence, choosing an action is the same thing as deciding to try to perform the action if you can try to perform it. So choosing an action amounts to adopting this conditional policy, and the policy can be evaluated by computing its expected value (or marginal expected value). This is the intuitive rationale for the definition of expected utility.

Note that if we adopt the "abstract" reading of classical decision theory according to which it is left open what count as decision-theoretic actions, then one way of understanding the present proposal is that the optimality prescription yields the correct prescriptions if it is restricted to these conditional policies.

## 6. Two Problems

At this point let me address two general problems for the evaluation of actions in terms of trying and being able to try to perform them. When the first version of this account was published in Pollock (2003), Peter McInerney (2006) published an interesting response. He observed that we can try harder or less hard to do *A*, and how hard we try can affect both the execution costs and how likely we are to succeed. It seems that the definition of the expected utility of an action should take this into account. This is actually a special case of a more general phenomenon. Typically, high-level actions can be performed in many different ways. Suppose *A* is the action of *going to the store to buy an avocado*. I could walk, or ride a bike, or drive, or take a cab, or fly to Cleveland and hitchhike back to Tucson, and so on. These are all ways of performing *A*. But they have different execution costs and may have different probabilities of achieving desired outcomes. How then do we assign a unique expected utility to *A*?

The solution lies in the observation that the agent has to decide not just whether to perform *A*, but also how to perform *A*. As I observed in chapter 1, planning is typically schematic and hierarchical. We construct plans built out of high-level actions, and then we must plan further for how to perform those actions. If we are rational, we will try to find the best ways we can to perform the actions. In other words, when we perform *A* we will try to do so in a way that maximizes **EU**(*A*). So in deciding whether to perform *A*, we should assume that we will perform it in the best way we can. In other words, the value of **EU**(*A*) should be the maximum value for all the ways we can think of for performing *A*. This handles McInerney's objection too. In evaluating **EU**(*A*), we must assume that we will try however hard maximizes **EU**(*A*).

A further complication arises from the fact that we often make our initial decision about whether to perform *A* before we have time to plan exactly how we are going to perform *A*. In this case we must rely upon stored information about what it is generally like to perform *A*. When we get around to planning for how to perform *A*, we may be surprised to discover that in this case **EU**(*A*) is quite different from what we would generally expect, and this may necessitate our reconsidering our earlier decision.

A second problem for the present proposal is that it is sometimes unclear whether we are unable to try to perform the action or just unable to succeed in performing it. For example, imagine a person who knows that he cannot jump over the moon, but when offered some reward from trying to do so looks up at the moon and jumps as high as he can. Is he really trying, or is he just pretending to try? Or consider a person who wants to paint a block, but when he looks for the paint he is unable to find it. Does that prevent him from being able to try, or does looking for the paint count as trying to paint the block? People's intuitions differ on cases like this. However, it turns out that what we say about these cases makes no difference to the expected utility of the action. Let $C$ describe the cases in which it is agreed that the agent can try to $A$, and let $C^*$ be the cases in which we are undecided whether the agent is unable to try to perform $A$ or just unable to succeed. So in computing $\textbf{C-PROB}_{try\text{-}A\ if\ can\text{-}try\text{-}A}(O)$ we are undecided whether it should be

$$\textbf{C-PROB}_{try\text{-}A\ if(C \vee C^*)}(O)$$

or

$$\textbf{C-PROB}_{try\text{-}A\ if\ C}(O).$$

$C$ and $C^*$ are supposed to describe different cases, so $C$ entails $\sim C^*$. The outcomes $O$ we are interested in will normally be such that even if we grant that the agent can try to perform $A$ in the cases described by $C^*$, doing so will be inefficacious in bringing about $O$. For example, trying to paint the block but not being able to find the paint will have no effect on getting the block painted. In that case, $\textbf{C-PROB}_A(O/C^*) = \textbf{PROB}(O/C^*)$, and we can then prove that $\textbf{C-PROB}_{try\text{-}A\ if\ can\text{-}try\text{-}A}(O)$ is unaffected by whether we include $C^*$ as cases in which the agent can try. This results from the following theorem:

**Theorem 6:** If $C$ entails $\sim C^*$ and $\textbf{C-PROB}_A(O/C^*) = \textbf{PROB}(O/C^*)$ then $\textbf{C-PROB}_{A\ if(C \vee C^*)}(O) = \textbf{C-PROB}_{A\ if\ C}(O)$.

Proof: First note that

$$\textbf{PROB}(\sim C)\cdot\textbf{PROB}(O/\sim C)$$

$$= \textbf{PROB}(\sim C)\cdot[\textbf{PROB}(O/C^*)\cdot\textbf{PROB}(C^*/\sim C) + \textbf{PROB}(O/\sim C \& \sim C^*)\cdot\textbf{PROB}(\sim C^*/\sim C)]$$

$$= \textbf{PROB}(O/C^*)\cdot\textbf{PROB}(C^*) + \textbf{PROB}(O/\sim C \& \sim C^*)\cdot\textbf{PROB}(O/\sim C \& \sim C^*).$$

Then

$$\textbf{C-PROB}_{A\ if(C \vee C^*)}(O)$$

$$= \mathbf{PROB}(C \vee C^*) \cdot \mathbf{C\text{-}PROB}_A(O/C \vee C^*) + \mathbf{PROB}(\sim C \& \sim C^*) \cdot \mathbf{PROB}(O/\sim C \& \sim C^*)$$

$$= \mathbf{PROB}(C \vee C^*) \cdot \mathbf{PROB}(C/C \vee C^*) \cdot \mathbf{C\text{-}PROB}_A(O/C) + \mathbf{PROB}(C \vee C^*) \cdot \mathbf{PROB}(C^*/C \vee C^*) \cdot \mathbf{C\text{-}PROB}_A(O/C^*) + \mathbf{PROB}(\sim C \& \sim C^*) \cdot \mathbf{PROB}(O/\sim C \& \sim C^*)$$

$$= \mathbf{PROB}(C) \cdot \mathbf{C\text{-}PROB}_A(O/C) + \mathbf{PROB}(C^*) \cdot \mathbf{PROB}(O/C^*) + \mathbf{PROB}(\sim C \& \sim C^*) \cdot \mathbf{PROB}(O/\sim C \& \sim C^*)$$

$$= \mathbf{PROB}(C) \cdot \mathbf{C\text{-}PROB}_A(O/C) + \mathbf{PROB}(\sim C) \cdot \mathbf{PROB}(O/\sim C)$$

$$= \mathbf{C\text{-}PROB}_{A \text{ } if \text{ } C}(O). \quad \blacksquare$$

The upshot is that for decision-theoretic purposes, it makes no difference how these intuitively problematic cases are resolved.

## 7. Computing Expected Utilities

I have shown how to modify classical decision theory to accommodate the failure of action omnipotence. This is supposed to be part of a theory of real rationality (as opposed to ideal rationality). Is it plausible to suppose that real agents, with realistic cognitive resource constraints, can actually compute expected utilities?

Expected utilities are defined in terms of expected values, and as noted above, the expected value of an action is the expected value of the entire world when the action is performed. Clearly, that is not something that we can compute. However, $\mathbf{EU}(A) = \mathbf{PROB}(\textit{can-try-A}) \cdot \mathbf{MEV}(\textit{try-A}/\textit{can-try-A})$. Thus the agent doesn't have to know the value of $\mathbf{EV}(\textit{try-A}/\textit{can-try-A})$. All the agent has to know is the conditional *marginal* expected value of trying to perform each action—the *difference* between the conditional expected value of trying to perform the action and the expected value of the null-action. In computing this, we can ignore those aspects of the world that are unaffected by the choice between the actions. All that is relevant to the conditional marginal expected value of an action is those value-laden aspects of the world that are affected by the choice.

The problem may still seem impractically difficult. How can we know all the value-laden parameters that are causally affected by an action? But this is not so difficult as it sounds. By virtue of the principle of irrelevance discussed in chapter 5, it is defeasibly reasonable to regard a parameter as value-neutral unless we have some reason for thinking otherwise. Thus we can ignore most parameters. Of the parameters that are not value-neutral, the principles of direct inference discussed in chapters 7 and 8 make it defeasibly reasonable to assume that the probability of any particular one is not causally affected by the action unless we have some reason for thinking otherwise. Thus it is defeasibly reasonable to ignore all aspects of the world except those that we have reason to believe to be value-laden and probabilistically affected by the action, and that narrows the computation of expected utilities down to a manageable set of parameters.

The upshot of this is that the computational problem of computing expected utilities when the possible outcomes are entire possible worlds has a solution. We will not be able to compute expected values, but for computing expected utilities we do not have to compute expected values. We need only compute marginal expected values, and that is made manageable by principles for reasoning defeasibly about both values and probabilities.

## 8. Conclusions

A theory of rational choice is a theory of how an agent should, rationally, go about deciding what actions to perform at any given time. This is a theory about choosing between ordinary actions. If we adopt the optimality prescription as our theory of rational choice, taking the actions to which it applies to be ordinary actions, we find that it can yield intuitively incorrect decisions by ignoring the fact that it is sometimes difficult or impossible to perform an action and sometimes one cannot even to try to perform it. One way to attempt to repair the theory is to restrict it to a class of actions that can be performed infallibly, but there does not seem to be any appropriate class of actions having this property. The only other alternative appears to be to revise the definition of expected value to take account of the failure of action omnipotence. That led us to the rather ad hoc looking definition of "expected utility". However, that definition can be justified by the observation that comparing actions in terms of their expected utilities is equivalent to comparing them in terms of the expected values of the associated conditional policies of the form *try-A if can-try-A*. In light of the failure of action omnipotence, choosing an action is tantamount to adopting such a conditional policy, and so the evaluation of the action should be the same as the evaluation of the conditional policy.

Thus far we have considered two problems for the optimality prescription, and we have shown how to meet them by modifying the optimality prescription in fairly conservative ways. Unfortunately, more problems remain. In the next chapter we will encounter difficulties that cannot be met by making minor changes to the theory. They will force the abandonment of the optimality prescription and its replacement by a different kind of theory of rational choice.

# 10
# Plans and Decisions

## 1. Against Optimality

Thus far we have investigated two kinds of problems for the optimality prescription. The first concerned the "informational" nature of mixed physical/epistemic probabilities, and was easily avoided by reformulating the definition of expected values in terms of causal probabilities. The second concerned the failure of action omnipotence. This required more drastic surgery to the optimality prescription. We found that either we had to radically revise the criterion for evaluating actions, using expected utility in place of expected value, or we had to take the objects of decision-theoretic evaluation to be conditional policies rather than ordinary actions. Still, with those changes we were left with a principle having the same form as the optimality prescription. In this chapter we turn to difficulties that will finally do in the optimality prescription altogether. These turn upon the fact that actions cannot be evaluated or chosen in isolation. In the general case, the proper objects of decision-theoretic evaluation turn out to be plans (courses of action) rather than individual actions. Although plans are evaluated in terms of their expected utilities, they cannot be chosen on the basis of a pairwise comparison because plans can differ in scope. I propose an alternative picture, called "Locally Global Planning", according to which plans are evaluated in terms of their contribution to the agent's "master plan". However, when engaging in locally global planning, there are always potentially infinitely many plans that compete with a given plan. An agent cannot be expected to survey them all, so it is unrealistic to expect the agent to choose optimal plans. There may not even be optimal plans. For every plan, there may be a better plan. So there is no way to save the optimality principle. The objective of decision-theoretic deliberation cannot be to find optimal solutions to practical problems. The agent must instead search for good solutions, and when better solutions arise, use them to replace solutions adopted earlier.

> The focus of this book is on the theory of practical decision making. However, it has direct implications for AI planning theory, a field in which I also work. Throughout this chapter I will include indented parenthetical remarks aimed at AI planning theory. When I don my AI hat, my objective is to implement a decision-theoretic planner that makes rationally correct decisions. But before we can do that, we must have an account of rational decision making that is applicable to this problem.

Decision-theoretic planners typically assume "simple plan-based decision theory", which results from applying the optimality prescription to plans rather than actions. I am in the process of arguing that classical decision theory is not a correct theory of rational decision making, and in this chapter I will argue that simple plan-based decision theory is not correct either. So all current decision-theoretic planners are based on an incorrect theory of rational decision making. There is a theoretical problem here that must be solved before we can even begin to implement. It will follow from the conclusions drawn in this chapter that all existing planners will produce logically incorrect solutions when applied to the complex planning problems facing a sophisticated autonomous agent operating in an environment of real world complexity (if they are able to produce any solutions at all). This is important. It means that planning theorists must look in new directions to construct a generally adequate planner.

I could stop there and claim to have shown something important. However, within AI my ultimate objective is to produce an alternative planning system that will produce logically correct solutions to these planning problems, and that requires the theory to be worked out in further detail. The problem of completing the theory and constructing the planner go hand in hand. The theory cannot be correct if it cannot be implemented, so some aspects of the theory will be driven by thinking about implementation. The final chapter of this book looks at these issues in a preliminary way. It is unlikely that a fully adequate theory will be constructed without actually doing the implementation. Thus this book represents only a preliminary attack on that problem. Getting the outlines of the theory right is the necessary first stage in constructing a successful decision-theoretic planner, and it must be done before one can even attempt to implement. I feel that it is important to lay the theory out as best we can at this point. This is a necessary step before we can make further progress.

## 2. The Logical Structure of Practical Deliberation

### 2.1 *Deciding-Whether and Deciding-Which*

The fundamental problem of practical deliberation is what to do. Such deliberation comes in two forms. Sometimes we are deciding whether to perform a particular action. At other times we are deciding which of several actions to perform. A fundamental presupposition of decision theory is that deciding-whether reduces to deciding-which. Deciding-which is a matter of deciding between alternatives, and the alternatives are to be evaluated, in accordance with the optimality prescription, in terms of their expected utilities.

There is a trivial sense in which deciding-whether is a case of deciding which. In deciding whether to do *A*, we are deciding between doing *A* and not doing *A*. Let $\overline{A}$ be the action of *not* doing *A*. Then deciding whether to do *A* is a matter of deciding which of *A* and $\overline{A}$ to do. However, this decision cannot be made by comparing the expected utilities of *A* and $\overline{A}$. Jeffrey (1965) considers the example of deciding whether to bring red wine or white wine to a dinner party. This is an example of deciding-which. Suppose instead that we just consider whether we should bring red wine. Let this be the action $A$. Can we make this decision by comparing the expected utility of $A$ with the expected utility of $\overline{A}$? No. In computing the expected utility of $\overline{A}$, we cannot assume that if we do not take a bottle of red wine then we will take a bottle of white wine. We might take nothing at all. There is no way to predict that we will take a bottle of white wine until we have solved the decision problem at least to the extent of determining that these are the only two viable alternatives. Thus the expected utility of $\overline{A}$ will be essentially 0. Hence applying the optimality prescription in this way would lead us to take a bottle of red wine. However, by the same reasoning, it is better to take a bottle of white wine than not to take a bottle of white wine, so this reasoning will also lead us to take a bottle of white wine. And we do not want to take both. Thus we cannot choose between $A$ and $\overline{A}$ in this way.

Decision theory assumes that in deciding whether to perform $A$, we should consider what we might do instead, where this is more than just not doing $A$. Thus in the wine case, we evaluate the action of taking red wine by comparing it with the action of taking white wine. These are considered "alternatives", and the assumption is that we should choose $A$ only if there are no better alternatives. But this cries out for an explanation of what counts as an alternative. The optimality prescription is formulated in terms of a given set of alternative actions. It tells us how to choose between the alternatives, but it doesn't tell us how to determine what actions are alternatives. Without that, the prescription cannot be applied to real cases and does not constitute a theory of rational choice.

It will be my contention that this entire approach to rational decision making is fundamentally misguided. Deciding-whether questions cannot, in general, be reduced in this way to deciding-which questions. This will turn upon the claim that there is, in general, no way to make sense of the notion of a decision-theoretic alternative to an action so that the optimality prescription constitutes a correct theory of rational decision making. My argument for this claim will be rather involved. I will argue first that we cannot evaluate actions in isolation—they must be evaluated in combination with other actions, that is, as parts of plans. Then I will argue that there is no way to select a fixed set of alternatives from the potentially infinite array of plans available, and even if there were there would be no reason to expect there to be optimal plans. Finally, I will argue that instead of looking for optimal plans, we should be looking for good plans, and I will attempt to make that notion precise.

### 2.2 *Strong and Weak Competition*

It might be supposed that the search for a set of alternatives has a trivial solution—any set of actions can constitute the alternatives, it is up to the decision maker. The alternative actions are just those the decision maker is considering performing. There are two obvious problems for this proposal. The first is that the decision maker may not be considering everything he should be considering. If I employ the optimality prescription to decide to take red wine without even considering the possibility of taking white wine, it does not follow that my decision is the right one. A second problem is that at any given time a decision maker may be considering many different actions and may choose to perform more than one of them. I may be considering going to lunch at noon, and reading a novel this evening. There is no reason why I should not do both. These actions are not in competition with each other.

The first problem is just the problem of finding the set of alternatives, but the second problem is more fundamental Apparently the actions in a set of alternatives must be, in some sense, "competing actions" that I must choose between. But what is it for actions to compete? It is frequently supposed that competing actions are those that cannot be performed together. Let us call these *strongly competing* actions. This is often (but not always[1]) built into the formal definition of a decision problem. It is often required that **A** is a set of actions that are pairwise logically incompatible (it is logically impossible to perform more than one of them) and exhaustive (it is logically necessary that you will perform at least one of them). In other words, **A** is a "partition of the action space".

However, commonsense decision problems do not generally involve choices between strongly competing actions. In Jeffrey's example of deciding whether to take red wine or white wine to a dinner party. One could, of course, take both. These actions do not strongly compete. In fact, taking our ordinary descriptions of our decisions at face value, choices between strongly competing actions seem to be the exception rather than the rule. I face such decision problems as deciding whether to paint the shed this afternoon or clean the house, whether to fly to LA next week or the following week, whether to cook chicken for dinner or cook lamb chops, and so on. In all of these cases I could do both. Still, in each case the two choices are in some sense in competition. I do not regard doing both as a serious option. We want alternatives to be actions we should, rationally, choose between. That is, we should choose one but not more than one. This can be the result of much weaker relations than strong competition. For instance, in Jeffrey's wine example, we must choose between red wine and white wine because taking both would be foolish. The cost would be twice as great but the payoff only marginally greater, so the expected utility of taking both is less than the expected utility of taking one. Thus rather than take both, we should take one, but that necessitates deciding which to take. We might capture this with a notion of weak competition—two actions *compete weakly* iff either they compete strongly or the expected utility of doing both is less than the expected utility of at least one of them.

An appeal to weak competition generates a theory with a somewhat different structure than the optimality prescription. The problem is that the optimality prescription assumes we have a set of alternative actions, and prescribes choosing an optimal member of the set. However, weak competi-

---

[1] For instance, see Jeffrey (1983). It is also worth noting that in the literature on decision-theoretic planning, with the exception of MDP planning, the alternatives are never strong competitors.

tion doesn't generate a set of alternatives. This is because weak competition is not transitive. *Action*$_1$ may compete weakly with *action*$_2$, and *action*$_2$ with *action*$_3$, without *action*$_1$ competing weakly with *action*$_3$. Thus if we simply pick an action and let the set of alternatives be the set of all actions competing weakly with the given action, it does not follow that other members of the set of alternatives will be in competition. It may be desirable to perform several of those alternatives together. For instance, suppose I am planning a wedding. Folk wisdom dictates that I should select something borrowed and something blue, but suppose it is undesirable to select two borrowed things or two blue things. If $x$ and $y$ are borrowed, and $y$ and $z$ are blue, then selecting $x$ competes weakly with selecting $y$, and selecting $y$ competes weakly with selecting $z$, but selecting $x$ does not compete weakly with selecting $z$.

This problem does not depend upon taking weak competition as our competition relation. For instance, strong competition is not transitive either. However competition is to be defined, it seems that what the optimality prescription really ought to say is:

(OP) It is rational to decide to perform an action iff it has no competitor with a higher expected utility.

This is equivalent to talking about an optimal member of a set of alternatives only if the competition relation is transitive. There is no obvious reason to expect that to be the case, so I will henceforth assume that the optimality prescription takes the form of (OP).

### 2.3 Boolean Combinations

There is a way of reformulating these decision problems so as to obtain a set of pairwise strongly competing alternatives. Where $A_1,...,A_n$ are the actions we are choosing between, a *Boolean combination* of them is a specification of which $A_i$'s to perform and which to not perform. For example, where $\overline{A}$ is the action of *not* performing $A$, a Boolean combination might have the form $A_1$ & $\overline{A}_2$ & $A_3$ & ... . Different Boolean combinations are logically incompatible with each other, and the disjunction of them is logically necessary.[2] The actions $A_1,...,A_n$ compete weakly just in case no Boolean combination with multiple positive constituents has as high an expected utility as some Boolean combination with just one positive constituent. Thus we can retain the original form of the optimality prescription if we apply it to the Boolean combinations of what are intuitively "alternative actions".

Although the appeal to Boolean combinations allows us to use strong competition as our competition relation, it is not clear that it really buys us anything. The problem is that the negative elements of a Boolean combination do not appear to make any contribution to the probabilities involved in computing expected utilities. When we discussed the comparison of $A$ with

---

[2] Note that we can identify the null action (discussed in chapter 9) with $\overline{A}_1$ & ... & $\overline{A}_n$.

$\overline{A}$, we noted that supposing we do not perform $A$ leaves it pretty much wide open what actions we do perform and what else will happen in the world. That seems to be equally true for Boolean combinations in general. Supposing, for example, $A_1$ & $\overline{A}_2$ & $A_3$, will not usually give us any more useful information than just supposing $A_1$ & $A_3$. If this is right then applying (OP) to the individual actions is equivalent to selecting the optimal Boolean conjunction, but the fact that we get strong competition by looking at the Boolean combinations is irrelevant to the decision making. We get the same result without strong competition by just considering actions that compete weakly.

### 2.4 *Universal Plans*

Regardless of whether we understand competition in terms of weak competition, or strong competition, or something else, we still need an account of what competing alternatives we must compare an action to. For example, on any account of competition it will presumably be the case that $A$ and $\overline{A}$ are competitors, but as we have seen, in deciding whether to perform $A$ it is not sufficient to just compare its expected utility with that of $\overline{A}$. Conversely, even if $B$ is not, intuitively, a competitor of $A$, $\overline{A}$&$B$ competes strongly with $A$ and, as we have seen, the expected utility of $\overline{A}$&$B$ will normally be the same as the expected utility of $B$. In deciding whether to take red wine, we do not want the expected utility of that to have to surpass the expected utility of every unrelated action, like vacationing in Brazil next summer, but if $\overline{A}$&$B$ were regarded as a competitor, it would have that effect. Thus there has to be something more to the set of alternatives than that it contains actions that compete with $A$. How should we select the set of alternatives?

One possibility is that all actions and combinations of actions should be regarded as potential competitors. Notice that the possible alternatives include more than just actions to be performed at the same time. For instance, I may want to choose between flying to LA next week or flying the following week. This suggests that **A** should consist of the Boolean combinations of all possible actions. On this approach, all actions, for all times, must be considered together in the Boolean combinations. A Boolean combination becomes a complete prescription of what to do for the rest of one's life, and a decision problem is reduced to the problem of finding the optimal such course of action. We might call these "universal plans".

Savage (1954) toys with the idea that rational decisions should be between universal plans, but he rejects it for the obvious reason. It is computationally absurd to suppose I must plan the entire rest of my life in order to decide what to have for lunch. The real world is too complex for us to be able to construct or compare universal plans. No decision maker with realistic computational limitations could possibly govern its life by finding a plan prescribing the optimal action for every instant of the rest of its life. Most likely such plans would involve infinitely many individual actions (including all the not-doings), and finding an optimal plan would involve comparing infinitely many alternatives. Even if the agent only has to consider a finite number $n$ of "atomic" actions, the number of Boolean combinations will be $2^n$, and that number quickly becomes immense. Even for a number of actions that is implausibly small for describing an entire life (for example, 300), $2^n$ will be a much larger number of Boolean combinations than a real agent can possibly survey and compare.

As I have repeatedly emphasized, the topic of this book is rational decision making by real agents, not ideal agents. We seek a theory of how a real decision maker should, rationally, go about deciding what actions to perform at any given time. A theory that requires a decision maker to do something that is impossible cannot be a correct theory of rationality. Real decision makers cannot construct and evaluate universal plans, so the theory of rationality cannot require them to.

Surprisingly, there is a strain of AI planning theory that aims at constructing universal plans of precisely the sort I have been disparaging. This is Markov decision process (MDP) planning.[3] It is to be emphasized that the preceding remarks are only aimed at agents with fairly sophisticated aspirations. If one wants to build a very simple planning agent that is only able to perform a very narrow range of tasks, then one might solve the problem by being very selective about the properties considered, and the agent might be able to construct a universal plan. This might work for a mail delivery robot, but it cannot possibly work for an agent as complex as a planetary explorer — or at least, not a good one. And it is interesting to see how easily the problem can arise even in quite restrictive domains. Consider the following problem, which generalizes Kushmerick, Hanks and Weld's (1995) "slippery gripper" problem. We are presented with a table on which there are 300 numbered blocks, and a panel of correspondingly numbered buttons. Pushing a button activates a robot arm which attempts to pick up the corresponding block and remove it from the table. We get 100 dollars for each block that is removed. Pushing a button costs two dollars. The hitch is that half of the blocks are greasy. If a block is not greasy, pushing the button will result in its being removed from the table with probability 1.0, but if it is greasy the probability is only 0.1. We are given exactly 300 opportunities to either push a button or do nothing. Between button pushes, we are given the opportunity to look at the table, which costs one dollar. Looking will reveal what blocks are still on the table, but will not reveal directly whether a block is greasy. What should we do? Humans find this problem terribly easy. Everyone I have tried this upon has quickly produced the optimal plan: push each button once, and don't bother to look at the table.

We can cast this as a 599 step POMDP (partially observable Markov decision problem). Odd-numbered steps consist of either pushing a button or performing the null action, and even-numbered-steps consist of either looking at the table or performing the null action. World-states are determined by which blocks are on the table ($T_i$) and which blocks are greasy ($G_i$). The actions available are *nil* (the null action), $P_i$ (push button $i$), and $L$ (look at the table). This cannot be cast as a FOMDP (fully observable Markov decision problem), because the agent cannot

[3] See Boutelier et al. (1999) for a good survey of MDP planning.

observe which blocks are greasy. But notice that even if it could, we would immediately encounter an overwhelming computational difficulty. The number of world states would be $2^{600}$, which is approximately $10^{180}$. As noted earlier in the book, it has been estimated that total the number of elementary particles in the universe is approximately $10^{78}$. So the number of world-states is 102 orders of magnitude greater than the estimate of the total number of elementary particles in the universe. The state-space gets even larger when we move to POMDP's. In a POMDP, the agent's knowledge is represented by a probability distribution over the space of world-states. The probability distributions constitute *epistemic states*, and actions lead to transitions with various probabilities from one epistemic state to another. In general, POMDP's will have infinite spaces of epistemic states corresponding to all possible probability distributions over the underlying state-space, however a reachability analysis can often produce a smaller state-space with just finitely many possible probability distributions. In the slippery blocks problem, it can be shown that in reachable epistemic states **PROB**($T_i$) can taken any value in the set {1.0, $.5{\cdot}.9$, $.5{\cdot}.9^2$, ... , $.5{\cdot}.9^{300}$, 0} and **PROB**($G_i$) can take the values .5 and 1.0. Not all combinations of these values are possible, but the number of reachable epistemic states is greater than $301^{300}$, which is approximately $10^{744}$. A universal plan is equivalent to a policy for this state space. There is no way to even encode such a policy explicitly.

On the other hand, as remarked above, humans have no difficulty solving this planning problem? Why? Part of the answer is that to find an optimal plan, we do not have to consider all of the states that could be reached by executing other plans. Most of those states are irrelevant to our objectives, and so there is no need to find (and no possibility of finding) a universal plan. And this arises already for what is basically a toy problem. The real world can be expected to be much more complex, even for agents that ignore most of it. So autonomous agents operating in the real world must aim at constructing local plans, not universal plans.

## 2.5 *The Search for Alternatives*

What the preceding considerations indicate is that we have to decide what the reasonable alternatives should be before we can apply the optimality prescription. We cannot take every action indiscriminately to be among the competitors (or constituents of competitors). We must be more discriminating in our selection of alternatives. We need something in between taking $\mathbf{A}$ to consist of just $A$ and $\overline{A}$ and taking it to consist of all possible universal plans. How do we accomplish that? The optimality prescription itself is no help in answering this question. But without an answer to this question, we do not have a theory of rational choice.

I have suggested that the appeal to Boolean combinations is just a technical trick that does not accomplish anything we cannot achieve more simply by appealing to weak competition and applying (OP). So I will frame my subsequent discussion around (OP), although nothing really turns on that. The question remains, which weakly competing actions must be considered alternatives in order to make our choices rational? I do not see any plausible way to answer this question, so let us set it aside. In the next section, I will give reasons for thinking that the optimality prescription would not work even if we could answer this question.

## 3. Groups of Actions

The principle (OP) evaluates actions by comparing them with other individual actions. The next step of my attack on the optimality prescription is an argument that we cannot, in general, make decisions by focusing on individual actions. I will argue that the proper objects of decision-theoretic evaluation are plans rather than individual actions. At this stage, we can give at least three reasons why this must be the case.

The first reason turns upon the observation that neither weak competition nor any other reasonable competition relation can be expected to be transitive. For example, in the "borrowed and blue" example, suppose choosing *z* (a blue thing) has a higher expected utility than choosing *y* (a borrowed and blue thing), and choosing *y* has a higher expected utility than choosing *x* (a borrowed thing). Thus (OP) implies that one ought to choose *z* (a blue thing), but it also implies that it is not reasonable to choose either *x* or *y* because both have competitors with higher expected utilities. Then we are left without a borrowed thing. On the contrary, it seems clear that if we choose *z* as our blue thing, then because we want to avoid choosing two blue things, we should choose *x* as our borrowed thing. To get this result, we must consider choosing *x* and *z* as a package, and compare that with choosing *y* alone. So we cannot evaluate *x* in isolation. We have to look at groups of actions rather than single actions.

We can be led to this same conclusion by reflecting on the fact that we typically have a number of different decisions to make at more or less the same time. I may be deciding whether to go to the bank before lunch or after lunch, and also deciding where to go for lunch. This mundane observation again creates a problem for the optimality prescription because (OP) evaluates actions one at a time and has us choose them individually on the basis of their being optimal. The problem is that decisions can interact. Carrying out one decision may alter the probabilities and utilities involved in another decision, thereby changing what action is optimal. It could be that, prior to deciding where to go to lunch, because I am very hungry the optimal decision would be to postpone going to the bank until after lunch. But if I decide to have lunch at a restaurant far from the bank and I have other things to do in that part of town that could occupy me for the rest of the afternoon, this may make it better to go to the bank before lunch. Alternatively, because I am very hungry and want to eat before going to the bank, it might be better to choose a different restaurant. The point is that actions can interfere with one another, with the result that if several actions are to be chosen, their being individually optimal does not guarantee that the group of them will be optimal. This strongly suggests that the object of decision-theoretic evaluation should be the entire group of actions rather than the individual actions.

This same conclusion can be defended in a third way. Often, the best way to achieve a goal is to perform several actions that achieve it "cooperatively". Performing the actions in isolation may achieve little of value. In this case we must choose actions in groups rather than individually. To illustrate, suppose we have a table suspended from the ceiling by cables, and it is laid with expensive glassware. We can raise the right side of the table by activating a servomotor that retracts the cable on the right, and we can raise the left side by activating a different servomotor. We want to raise the table. Activating the right servomotor by itself would raise only the right side of the table and so spill the glassware onto the floor. Thus it has a negative expected utility. Similarly for activating only the left servomotor. What we must do is activate both servomotors. That has a positive expected utility even though it is composed of two actions having negative expected utilities. This illustrates again that actions cannot always be considered in isolation. Sometimes decision-theoretic choices must be between groups of actions, and the performance of a single action becomes rational only because it is part of a group of actions whose choice is dictated by practical rationality.

The last two examples illustrate two different phenomena. In the first, actions interfere with each other, changing their execution costs and hence their expected utilities from what they would be in isolation. In the second, actions collaborate to achieve goals cooperatively, thus changing the expected utilities by changing the probabilities of outcomes. These examples might be viewed as cases in which it is unclear that actions even have well-defined expected utilities in isolation. To compute the expected utility of an action we must take account of the context in which it occurs. If the expected utilities are not well-defined, then the optimality prescription cannot be applied to these decision problems. Alternatively, if we suppose that the expected utilities of the actions in isolation are well-defined, then what is important about these examples is that in each case we cannot choose the group of actions by choosing the individual actions in the group on the basis of their expected utilities. In the first example, the expected utility of the group cannot be computed by summing the expected utilities of the actions in the group. In the second example, the members of the group would not be chosen individually on their own strength. In these examples, it is the group itself that should be the object of rational choice, and the individual actions are only derivatively rational, by being contained in the rationally chosen group of actions.

Groups of actions, viewed as unified objects of rational deliberation, are *plans*. The simplest plans are linear sequences of actions. In general, plans can be viewed as "recipes" for action. I will say more about the logical structure of plans below. For now, the important point is that the actions in a plan may be good actions to perform only because they are part of a good plan. It appears that the only way to get the optimality prescription to make the right prescription in the above examples is to apply it to plans rather than individual actions. For instance, the reason we should activate the servomotor the lifts the right side of the table is that doing so is part of the plan of activating both servomotors, and the latter plan has a high expected utility.

Notice that in all three examples, if we apply the optimality prescription to Boolean combinations of actions rather than individual actions, we get the right answer. This is because the Boolean combinations are themselves plans. However, plans need not have the form of Boolean combinations. This is because (1) by and large, plans do not tell us what *not* to do, just what to do, and (2) plans can have more complicated logical structures than Boolean combinations. For example, plans can contain conditional steps telling us to do something only if something else is the case. Consider two route plans. One might say, "Take Speedway unless you encounter road construction. If you do encounter road construction, take Grant instead." The second plan might say, "Take Speedway unless you encounter road construction. If you do encounter road construction, take Broadway instead." Such plans can prescribe different courses of action in some circumstances (if you encounter road construction) but not in others (if you do not encounter road construction), so although they are intuitively competitors, they are not strong competitors.

Traditionally, choices were supposed to be between individual actions, but now we have seen that rational choices must often be made instead between plans, and the individual actions in the plans become only derivatively rational by being prescribed by a rationally chosen plan. How then do we choose between plans? The obvious proposal is to simply apply the optimality prescription to plans rather than actions. *Simple plan-based decision theory* would propose that we choose between competing plans in terms of their expected utilities. Savage (1954) seems to suggest that plans can be chosen in this way, and most work on decision-theoretic planning in AI is based upon this idea (for example, see Blythe and Veloso 1997; Boutelier et al. 1999; Haddawy and Hanks 1990; Ngo, Haddawy and Nguyen 1998; Onder and Pollack 1997, 1999; Onder, Pollack and Horty 1998; Williamson and Hanks 1994).

At this point it should be obvious that simple plan-based decision theory still faces a glaring difficulty. It does not even address the issue of how we should pick the set of alternative plans with which we compare a given plan in deciding whether to adopt it. We might go the way of (OP) (for single actions) and propose that a plan should be compared with all possible plans with which it competes weakly. But I will argue in section 5 that (1) it is computationally impossible for a real agent to make this comparison, and (2) it would not yield the intuitively correct decisions anyway. To salvage plan-based decision theory we must find a different way to select the alternatives to which a plan is to be compared. We cannot just compare a plan with all the weakly competing plans. Furthermore, I will argue that we cannot compare the alternative plans just by comparing their expected utilities.

## 4. Actions and Plans

Before pursuing the details of the proposal that we extend the optimality prescription by applying it to plans, a digression is in order. On a sophisticated view of actions, the optimality prescription is already committed to evaluating plans decision-theoretically. This is because the distinction between actions and plans is not a clean one. Actions can be very plan-like. Consider the action of *making a cup of tea*. How do you make a cup of tea? Let us say that it consists of the following: (1) putting water in the kettle, (2) heating the water, (3) retrieving a tea bag, (4) placing the tea bag in a teapot, (5) pouring boiling water into the teapot, (6) letting the concoction sit for several minutes, (7) pouring the tea from the teapot into a cup. The latter is a plan. It is hard to see what the difference is between the action of *making a cup of tea* and this seven-step plan.

Recall that I have distinguished between acts and actions. *Acts* are individual spatio-temporally located performances. *Actions* are act-types. In rational decision making it is actions we are deliberating about. That is, we are deciding what type of act to perform. There is no way to deliberate about (or even think about) a particular merely possible act that is not performed.

I often perform one act *by* performing another. For example, I make a cup of tea by performing the sequence of acts described above. In turn, I put water in the kettle by picking it up, putting it under the tap, turning on the water, waiting until the water reaches the appropriate level in the pot, turning off the tap, and setting the kettle down on the counter. We can progress to lower and lower levels of acts in this way, but eventually we will reach acts like *grasp the handle of the teapot* or *raise my arm* that I can perform "directly", without performing them by performing some simpler act. A *basic act* is an act that is not performed by performing another act or sequence of acts.

When the act instantiating an action is not a basic act, you perform the action by doing something else. The "something else" is typically a sequence of lower-level actions, and that sequence constitutes a plan. Roughly following the terminology used in "hierarchical task decomposition planning" in the artificial intelligence literature, let us say that a plan *decomposes* an action iff one can perform the action by executing the plan. Perhaps the main difference between an action and a plan that decomposes it is that there is generally more than one way to perform the action. That is, actions can have multiple decomposition plans, allowing them to be performed in different ways on different occasions. In fact, some actions appear to have an unbounded set of decomposition plans. Consider the action of *traveling from Tucson to LA*. There are infinitely many ways of doing that if we consider all the different means of transportation and all the different routes that can be taken. Of course, on any given occasion we only perform the action by executing a single one of the decomposition plans.

The set of decomposition plans available for a particular action is open-ended. It is not fixed by the meaning of the action term. We can often invent new ways of performing an action. For example, if I want to turn on the light, and I am sitting down and unable to reach the light switch from my chair, but there is a ski pole on the table beside me, I might turn on the light by pushing the switch with the ski pole. This is a new decomposition plan that I have never previously considered.

What determines the range of possible decomposition plans for an action? The meaning of an action term often specifies the act type partly in terms of what it accomplishes. For example, the action of *making a cup of coffee* results in my having a cup of coffee. So does the action of *buying a cup of coffee*, so there is more to the action than the goal it aims at achieving. An action is also characterized in part by constraints on how its goal is to be achieved. *Making a cup of coffee* differs from *buying a cup of coffee* in that when I make a cup of coffee, I must brew it myself, whereas when I buy a cup of coffee I must acquire it by purchasing it from someone else. These are constraints on the decomposition plans, but the constraints are not usually sufficient to determine a unique decomposition plan. For example, I can make a cup of coffee using a percolator, or a French press, or pouring hot water into toddy syrup, and so on.

What this all suggests is that at least for an important class of actions, the identity of an action is determined by (1) a goal-type, and (2) a constraint on what plans for achieving goals of that type can constitute decomposition plans for the action. Very high-level actions, like *save the world*, impose virtually no constraints on their decomposition plans. They are characterized almost exclusively in terms of their goal-types. This is quite different from the way in which philosophers have traditionally thought of actions. They have viewed actions as being much more concrete, although that may in part reflect a conflation of acts and actions.

I defined a *potentially basic action* to be an act type that *can* be instantiated by a basic act. An example is *wiggle your finger*. Note, however, that potentially basic actions can typically be performed in nonbasic ways as well. For example, I can wiggle my finger by grasping it with my other hand and moving it up and down. You can perform a potentially basic action in either of two ways—by performing a basic act that instantiates the action, or by executing a decomposition plan for the action. For actions that are not potentially basic, you can only perform them by executing a decomposition plan. So for most actions, performing them consists of executing a decomposition plan. Clearly, the expected utility of trying to execute an action will be dependent upon *how* one tries to execute it, that is, on one's choice of a decomposition plan. I might fly from Tucson to LA via Phoenix or via Cincinnati. The execution costs of these decomposition plans will differ significantly, producing different expected utilities for the plans. So comparing the expected utilities of actions must always be done against some background assumptions about what decomposition plans may be employed. The decision maker need not have decided precisely what decomposition plan to use, but the range of decomposition plans under consideration must be sufficiently constrained to allow meaningful estimates of expected utilities. The expected utility of the action must be determined by the candidate decomposition plans. The only obvious way to do this is to compute expected utilities for the decomposition plans, and then identify the expected utility of the action with that of its candidate decomposition plans. If this is right, then choosing actions by comparing their expected utilities is the same thing as choosing decomposition plans by comparing their expected utilities. So it seems that a sophisticated view of actions commits the optimality prescription to comparing plans in terms of their expected utilities.

## 5. Choosing Between Plans

Just as for actions, we need not choose between plans unless they are in some sense in competition. If two plans are not in competition, we can simply adopt both. So to construct a plan-based theory of rational choice, we need an account of when plans compete in such a way that a rational choice should be made between them. Competing plans should be plans that we must choose between, rather than adopting both. A sufficient condition for this is that executing one of the plans makes it impossible to execute the other one. That is, the plans *compete strongly*. However, it is clear that we often want to choose between plans that compete in much weaker ways. For example, consider the two route plans mentioned in section 3. Because the plans have conditional steps, they can prescribe incompatible actions under some circumstances and not others. Then they must be viewed as competing, but they do not compete strongly. Just as for actions, we might try to capture this in terms of weak competition. Let us say that two plans *compete weakly* iff either they compete strongly or the plan that results from merging the two plans into a single plan has a lower expected utility than at least one of the original plans. It might be proposed, then, that two plans are competitors iff they compete weakly, and accordingly:

(PB) It is rational to adopt (decide to execute) a plan iff it has no weak competitor with a higher expected utility.

The proposal (PB) is formulated in terms of the expected utility of a plan. But how should that be defined? We were led to the concept of expected utility for actions by the failure of action omnipotence. It is even more obvious that we cannot be guaranteed of being able to execute a plan, or even of being able to try to execute it (the latter because we may be unable to try to perform some of the actions it prescribes). To see how the definition should go for plans, consider an action *A* having a fixed decomposition plan *p*. Deliberating about the action should be equivalent to deliberating about the plan. The expected utility of an action was defined as follows:

$$\mathbf{EU}(A) = \mathbf{MEV}(\textit{try-A if can-try-A}).$$

Performing action *A* will consist of executing its decomposition plan *p*. It seems to follow that trying to perform action *A* consists of trying to execute *p*. Thus

$$\mathbf{EU}(p) = \mathbf{MEV}(\textit{trying to execute p if the agent can try to execute p})$$
$$= \textsf{PROB}(\textit{the agent can try to execute p}) \cdot \mathbf{MEV}(\textit{trying to execute p / the agent can try to execute p}).$$

So I will adopt this as my definition of the expected utility of a plan. However, this definition does not yet tell us enough to enable us to actually compute the expected utilities of plans. What is it to try to execute a plan? I will return to that question in chapter 11 where I will give a more precise account of plans and their expected utilities. For now I will rely upon this definition and let the reader use his intuitions about how the expected utilities of plans should work.

To evaluate (PB), let us first reflect briefly on the nature of the plans a realistic decision maker must construct and evaluate. We have seen that they cannot be universal plans. They have to be plans of more limited scope. What is the nature of these less-than-universal plans? This is most easily understood by reflecting on the fact that, over the course of her life, a decision maker is not faced with a single fixed planning problem. First, her beliefs will change as she acquires experience of her environment and as she has time for further reasoning. This will affect what solutions are available for her planning problems. For instance, suppose our decision maker is hiking in the desert, and she learns that the cliffs above a certain trail have become infested with Africanized bees. In light of this newly discovered information, that trail must be avoided. Route plans that would have been acceptable prior to acquiring this new information cease to be acceptable in light of it. Second, as our decision maker acquires more knowledge of her environment, her goals may change. For example, if our hiker detects an approaching thunderstorm, she will adopt the goal of finding a sheltered place where she can wait out the storm. This is not a goal she had before. This is a "goal of attainment". An agent that engages in goal-directed planning will adopt new goals of attainment under various circumstances. Note that this is not just a matter of the agent's having a general goal like "Always avoid thunderstorms", because if on one occasion she is unable to take cover and weathers the storm out in the open, that general goal is henceforth unattainable no matter what she does, but that should not affect her trying to avoid future thunderstorms. This point is discussed in more detail in chapter 12. The upshot is that the planning problems she faces evolve over time. We cannot expect her to redo all of her previous planning each time she acquires new knowledge or new goals, so planning must produce lots of *local plans*. These are small plans of limited scope aiming at disparate goals.

If (PB) is to work, it must work when applied to local plans. Those are the kinds of plans that real decision makers construct and decide to adopt. However, there are two simple reasons why (PB) cannot possibly be correct when applied to local plans. The simplest reason is that there are infinitely many of them. Plans are logical entities of potentially unbounded complexity.

(PB) would have us survey and compare all possible local plans in order to determine whether they compete with a given plan and, if they do, to determine whether they have a higher expected utility. But this is an impossible task. No real agent can consider all possible competitors to a given plan, so he cannot make decisions in accordance with (PB).

The first problem is devastating enough, but it is worth noting that there is a second problem (taken from my 1992). Even if we could somehow survey and compare an infinite array of plans, (PB) would not yield rationally correct decisions. (PB) is simply wrong as a theory of rational choice. This arises from the fact that for any plan there will almost always exist a competing plan with a higher expected utility. To illustrate, suppose again that I am choosing between roasting chicken and barbecuing lamb chops for dinner. Suppose the former has the higher expected utility. This implies that the plan of barbecuing lamb chops for dinner is not rationally adoptable, but it does not imply that the plan of roasting chicken for dinner is adoptable, because some other plan with a higher expected utility may compete with it. And we can generally construct such a competing plan by simply adding steps to the earlier competing plan. For this purpose, we select the new steps so that they constitute a subplan aimed at achieving some valuable unrelated goal. For instance, we can consider the plan of barbecuing lamb chops for dinner and then later going to a movie. This plan still competes with the plan of roasting chicken for dinner, but it has a higher expected utility. Thus the plan of roasting chicken for dinner is not rationally adoptable. However, the competing plan is not rationally adoptable either, because it is trumped by the plan of roasting chicken for dinner and then later going to the same movie.

It seems clear that given two competing plans $P_1$ and $P_2$, if the expected utility of $P_1$ is greater than that of $P_2$, the comparison can generally be reversed by finding another plan $P_3$ that pursues unrelated goals and then merging $P_2$ and $P_3$ to form $P_2$+$P_3$. If $P_3$ is well chosen, this will have the result that $P_2$+$P_3$ still competes with $P_1$ and the expected utility of $P_2$+$P_3$ is higher than the expected utility of $P_1$. If this is always possible, then there are no optimal plans and simple plan-based decision theory implies that it is not rational to adopt any plan.

In an attempt to avoid this problem, it might be objected that $P_2$+$P_3$ is not an appropriate object of decision-theoretic choice, because it merges two unrelated plans. However, we often merge plans for unrelated goals. If I plan to run two errands (aimed at achieving two unrelated goals), and both errands require me to go in the same direction, I may merge the two plans by running both errands on a single trip.

The inescapable conclusion is that the rational adoptability of a plan cannot require that it have a higher expected utility than all its competitors. The problem is that plans can have rich structures and can pursue multiple goals, and as such they are indefinitely extendable. We can almost always construct competing plans with higher expected utilities by adding subplans pursuing new goals. Thus there is no appropriate set of alternatives to use in defining optimality, and hence no way to define optimality so that it is reasonable to expect there to be optimal plans. Consequently, simple plan-based decision-theory fails.

The failure of simple plan-based decision theory is of fundamental importance, so let me recapitulate the three reasons it fails. First, it is unlikely that there will, in general, be such things as optimal plans. Second, because plans cannot generally be compared just by comparing their expected utilities, optimality may not even be desirable. Finally, even in those unusual cases in which there are optimal plans and optimality is desirable, finding them will be computationally intractable. So deciding whether to adopt a plan cannot turn upon it's being optimal.

I have heard it suggested that this problem does not arise for state space planners, for example, Markov decision process planning. However, there are two ways of viewing state space planning. We could think of the entire world as a single state space and try to produce a single universal plan governing the agent's actions for all time. Universal plans are not extendable, so they cannot be trumped by more comprehensive plans. But as we have seen, the problem of finding universal plans is completely intractable. Alternatively, we might use state space planning techniques to find local plans by separating out small subproblems, but here the preceding difficulty recurs. You can compare the plans produced for a single subproblem in terms of their expected utilities, but if you expand the problem and consider more sources of value, more possible actions, and so on, in effect looking at a bigger toy problem, an optimal plan for the larger problem may not contain the optimal plan for the subproblem. So in this case there will usually be more comprehensive plans that trump the plans constructed for the simpler subproblems, and once again, no plan is ultimately adoptable. Hence the difficulty arises independently of the kind of planning algorithms employed.

## 6. AI Planning Theory: The Real World versus Toy Problems

AI planning theory has made an industry out of constructing planners that can solve small isolated planning problems, where only a small number of actions and environmental conditions are taken to be relevant. These have come to be called "toy problems". The earliest toy problems were "blocks world" problems. In these problems, the planner is faced with a table top and some children's blocks. The blocks are initially piled on top of each other in some specified configuration, and the planning problem is to figure out how to get them into a different configuration. Despite having limited scope, toy problems can get very difficult, and current planners are able to solve some impressively difficult problems. Furthermore, some of these toy problems can generate useful practical applications. However, the logic of toy problems is different from the logic of decision making in the real world. Typically, the search for an optimal plan in a toy problem can be made by searching a manageably small (finite or finitely describable) set of alternatives. In this case optimality may be well-defined, and the prescription to choose an optimal plan would seem sensible.

However, outside of toy problems, the supposition that we need only search a very limited range of plans is indefensible. Plans are mathematical or logical constructions of unbounded complexity. In the real world, if we can construct one plan for achieving a goal, we can typically construct infinitely many. For planning by real agents in the real world, talk of "planning domains" is inappropriate. The only planning domain is the whole world.

It should be emphasized that this is only a claim about autonomous agents operating in the real world—not in highly restricted artificial environments. Ultimately I want know how to build Isaac Asimov's *I-Daneel* and Arthur C. Clarke's *HAL*. I take it that this is one of the seminal aspirations of AI, although at this time no one is in a position to even try to do it. For now most would be delighted if they could build a robot with much more limited capabilities, for example, an autonomous planetary rover that can learn about its environment and meet the challenges it is apt to encounter. But even these more limited aspirations are subject to the problems discussed above. On the other hand, for some concrete applications it is possible to constrain the planner's environment sufficiently to turn the problem into what is in effect a toy problem. This may be possible if the planning problem can be fixed from inception and either (1) local plans aiming at distinct goals do not interact, or (2) it is possible to construct universal plans with manageably short time horizons and manageably few actions. In such a world, optimization will be both well-defined and desirable. But my point here is that the techniques that work for toy problems will, for purely logical reasons, not scale up to the problem of building an autonomous agent operating in the real world. For this difficulty to arise, the agents need not be all that sophisticated. The difference between toy problems and the real world is not just a difference of degree. Planning must work entirely differently. This is a logical problem that must be solved before we can even begin building a decision-theoretic planner of use in autonomous agents in the real world.

## 7. When Is a Plan a Good One?

The question we ultimately want to answer is, "How should a rational agent go about deciding whether a plan should be adopted?" There will be no point at which a decision maker has exhausted all possibilities in searching for plans. Despite this, decision makers must take action. They cannot wait for the end of a non-terminating search before deciding what to do, so their decisions about how to act must be directed by the best plans found to date —not by the best possible plans that *could* be found. The upshot is that plan adoption must be defeasible. Decision makers must work with the best knowledge currently available to them, and as new knowledge becomes available they may have to change some of their earlier decisions. If a better plan is found later, that should supplant the plan adopted initially.

For this account to work, we need a notion of "better plan" that can be used in deciding whether a plan should be adopted. I have argued that this cannot be cashed out as one plan merely having a higher expected utility than a second. To get a grip on this notion of one plan being better than another, let us begin by considering the limiting case in which a decision maker has no background of adopted plans, and a new plan is constructed. Should the new plan be adopted? The basic insight of the optimality prescription is that what makes a course of action (a plan) good is that it will, with various probabilities, bring about various value-laden states, and the cost of doing this will be less than the value of what is achieved. This can be assessed by computing the expected utility of the plan. In deciding whether to adopt the plan, all the decision maker can do is compare the new plan with the other options currently available to him. If this is the only plan the decision maker has constructed, there is only one other option—do nothing. So in this limiting case, we can evaluate the plan by simply comparing it with doing nothing.

Things become more complicated when one has already adopted a number of other plans. This is for two reasons. First, the new plan cannot be evaluated in isolation from the previously adopted plans. Trying to execute the previous plans may affect both the probabilities and the utilities employed in computing the expected utility of the new plan. For example, if the new plan calls for me to pick up my friend at the airport, the probability of my being able to do that may normally be fairly high. But if other plans I have adopted will result in my driving to Phoenix earlier in the day, then the probability of being able to retrieve my friend may be lower. So the probabilities can be affected by the context provided by my other plans. The same thing is true of the values of goals. Suppose the new plan is a plan for repairing my car. In the abstract, this may have a high value, but if it is performed in a context in which my other plans include replacing the car in a short while, the value of the repair may be seriously diminished. Execution costs can be similarly affected. If the new plan prescribes transporting an object from one location to another in a truck, this will be more costly if a previous plan moves the truck to the other side of town.

Clearly, the expected utility of the new plan must be computed "in the context of the decision maker's other plans". But what does that require? Roughly, the probabilities and utilities should be conditional on the situation the decision maker will be in as a result of having adopted and tried to execute parts of the other plans. However, there isn't just one possible situation the decision maker might be in, because the other plans will normally have their results only probabilistically.

The second reason it becomes more complicated to evaluate a new plan when the decision maker already has a background of adopted plans is that the new plan can affect the value of the old plans. If an old plan has a high probability of achieving a very valuable goal but the new plan makes the old plan unworkable, then the new plan should not be adopted. Note that this is not something that is revealed by just computing the expected utility of the new plan.

We have seen that normal planning processes produce local plans. How should the decision maker decide whether to adopt a new local plan? The decision must take account of both the effect of previously adopted plans on the new plan, and the effect of the new plan on previously adopted plans. We can capture these complexities in a precise and intuitively appealing way by defining the concept of the decision maker's *master plan*. This is the result of merging all of the plans the agent has adopted but not yet executed into a single plan.

Do not confuse the master plan with a universal plan. The master plan simply merges a number of local plans into a single plan. Each local plan talks about what to do under certain circumstances, so the resulting master plan talks about what to do under every circumstance mentioned by any of the individual local plans. But this is still a very small set of circumstances relative to the set of all possible world-states. If none of the local plans have anything to say about what to do in some new previously unconsidered situation, then the master plan doesn't either. But by definition, a universal plan must include a prescription for what to do in every situation. If we have $n$ local plans each making $m$ prescriptions of the form "If $C$ is true then do $A$", the master plan will contain $m{\cdot}n$ prescriptions. But supposing the conditions $C$ are all logically independent of each other, a universal plan for the state space generated by just this limited vocabulary will contain $2^{m \cdot n}$ prescriptions. Typically an agent will be capable of considering what to do in a much larger set of circumstances not yet addressed by any of the local plans it has adopted. If there are $N$ such circumstances, a universal plan must include $2^N$ prescriptions. For example, if the agent has thus far adopted 30 ten-step plans, the master plan will include 300 prescriptions, but a universal plan would have to consider at least $2^{300}$ (that is, $10^{90}$) prescriptions, and probably many orders of magnitude more.

Although master plans are totally different beasts from universal plans, they share an important property — master plans can be meaningfully compared in terms of their expected utilities. We can think of the master plan as the agent's tool for making the world better. The expected utility of the master plan is the agent's expectation of how good the world will be if he adopts that as his master plan. Thus one master plan is better than another iff it has a higher expected utility. Equivalently, rationality dictates that if an agent is choosing between two master plans, he should choose the one with the higher expected utility.

It may at first occur to one that the objective should be to find an optimal master plan. But that cannot be right, for two familiar reasons. First, it is unlikely that there will be optimal master plans that are smaller than universal plans. If a master plan leaves some choices undetermined, it is likely that we can improve upon it by adding decisions regarding those choices. But as we have seen, it is not possible for real agents to construct universal plans, so that cannot be required for rational choice. Second, even if there were optimal master plans, realistically resource-bounded agents could not be expected to find them. So rationality cannot require finding optimal master plans.

These points are fairly obvious, and yet they completely change the face of decision-theoretic reasoning. Planning and plan adoption must be done defeasibly, and actions must be chosen by reference to the current state of the decision maker's reasoning at the time he has to act rather than by appealing to the idealized but unreachable state that would result from the decision maker completing all possible reasoning and planning. Decision makers begin by finding good plans. The good plans are "good enough" to act upon, but given more time to reason, good plans might be supplanted by better plans.[4] The decision maker's master plan evolves over time, getting better and better, and the rules for rationality are rules directing that evolution, not rules for finding a mythical endpoint. We might put this by saying that a rational decision maker should be an *evolutionary planner*, not an optimizing planner.

## 8. Locally Global Planning

It appears that the aim of practical deliberation is to construct local plans and use them to improve the master plan. If the only way an agent had of finding a master plan with a higher expected value was to plan all over again from scratch and produce a new master plan essentially unrelated to its present master plan, the task would be formidable. Performing the requisite planning would at the very least be slow and complex, making it difficult for the agent to respond to emergency situations. And if the agent's master plan is sufficiently complex, the agent's inherent computational limitations may make the task impossible. It does not take a very large problem to bog down a planning procedure. The reader unfamiliar with the AI literature on planning may not appreciate the severity of this problem. A few years ago, the very best AI planners could solve toy problems described in terms of 53 independent variables, where the solution was a plan of 105 steps (Weld 1999). Typically, the master plan will be significantly larger than these toy problems. Furthermore, if every planning problem requires the construction of a new master plan, then every little planning problem becomes immensely difficult. To plan how to make a sandwich for lunch, I would have to replan my entire life.

Obviously, humans do not do this, and artificial agents shouldn't either. Normal planning processes produce local plans, not entire master plans. The only way resource-bounded agents can efficiently construct and improve upon master plans reflecting the complexity of the real world is by constructing or modifying them incrementally. When trying to improve his

---

[4] This is reminiscent of Herbert Simon's (1955) concept of "satisficing", but it is not the same. Satisficing consists of setting a threshold and accepting plans whose expected-utilities come up to the threshold. The present proposal requires instead that any plan with a positive expected utility is defeasibly acceptable, but only defeasibly. If a better plan is discovered, it should supplant the original one. Satisficing would have us remain content with the original.

master plan, rather than throwing it out and starting over from scratch, what an agent must do is try to improve it piecemeal, leaving the bulk of it intact at any given time. This is where local plans enter the picture. The significance of local plans is that they represent the building blocks for master plans. We construct master plans by constructing local plans and merging them together.

Earlier, we encountered the purely logical problem of how to evaluate a newly constructed local plan, given that we must take account both of its effect on the agent's other plans and the effect of the agent's other plans on the new plan. We are now in a position to propose a preliminary answer that question. The only significance of local plans is as constituents of the master plan. When a new local plan is constructed, what we want to know is whether the master plan can be improved by adding the local plan to it. Thus when a new plan is constructed, it can be evaluated in terms of its impact on the master plan. We merge it with the master plan, and see how that affects the expected utility of the master plan.

The upshot of all this is that a theory of rational choice becomes a theory of how to construct local plans and use them to systematically improve the global plan—the master plan. I call this *locally global planning*. As a first approximation, we might try to formulate locally global planning as follows. Let us define the *marginal expected utility* of the local plan $P$ to be the difference its addition makes to the master plan $M$:

$$\mathbf{MEU}(P,M) = \mathbf{EU}(M+P) - \mathbf{EU}(M).$$

If the marginal expected utility is positive, adding the local plan to the master plan improves the master plan, and so in that context the local plan is a good plan. Furthermore, if we are deciding which of two local plans to add to the master plan, the better one is the one that adds more value to the master plan. So viewed as potential additions to the master plan, local plans should be evaluated in terms of their marginal expected utilities, not in terms of their expected utilities simpliciter.

It is not quite accurate to say that it is rational to adopt a plan iff its marginal expected utility is positive. This is for two reasons. First, adding a new plan may only increase the expected utility of the master plan if we simultaneously delete conflicting plans. For example, suppose I have adopted the plan to barbecue lamb chops for dinner. Then I remember that I have chicken in the refrigerator, and so I construct the new plan of roasting chicken for dinner. I cannot improve the master plan by simply adding the latter local plan to it. That would result in my making two dinners but eating only one, and so would lower the expected utility of the master plan rather than raising it. To improve the master plan I must simultaneously delete the plan to barbecue lamb chops and add the plan to roast chicken.

Second, plans may have to be added in groups rather than individually. In general, a change to the master plan may consist of deleting several local plans and adding several others. Where $M$ is a master plan and $C$ a change, let $M\Delta C$ be the result of making the change to $M$. We can define the *marginal expected utility of a change* C to be the difference it makes to the expected utility of the master plan:

$$\mathbf{MEU}(C,M) = \mathbf{EU}(M\Delta C) - \mathbf{EU}(M).$$

The *principle of locally global planning* can then be formulated as follows:

**Locally Global Planning**
It is rational to make a change $C$ to the master plan $M$ iff the marginal expected utility of $C$ is positive, that is, iff $\mathbf{EU}(M\Delta C) > \mathbf{EU}(M)$.

This is my proposal for a theory of rational decision making. The theory has two parts: (1) it is rational to perform an action iff it is prescribed by a rationally adopted master plan; and (2) a master plan is adopted rationally iff it is the result of incremental updating in accordance with the principle of locally global planning. I propose this as a replacement for the optimality prescription. It captures the basic insight that rational decision makers should guide their activities by considering the probabilities and utilities of the results of their actions, and it accommodates the observation that actions must often be selected as parts of plans and the observation that optimality cannot be defined in such a way that practical deliberation can be viewed as a search for optimal solutions to practical problems. A decision maker should be an evolutionary planner, not an optimizing planner. The principle of locally global planning tells us how evolutionary planning should work. It involves a fundamental change of perspective from prior approaches to rational choice, because decision making becomes a non-terminating process without a precise target rather than a terminating search for an optimal solution.

## 9. Conclusions

I have argued that the optimality prescription founders on an uncritical appeal to alternatives. The optimality prescription would only be reasonable if rational choices were made from small precompiled sets of alternatives. Pursuing the question of what makes actions alternatives led us to the more fundamental observation that, in general, actions cannot be chosen in isolation. Actions can both interfere with each other, and collaborate to achieve goals cooperatively. To accommodate this, actions must be chosen as parts of plans.

We cannot save the optimality prescription by adopting a simple plan-based decision theory according to which it is rational to adopt a plan iff it is an optimal plan from a set of alternatives. The problems for simple plan-based decision theory are twofold. First, if we confine our attention to local plans, there is no apparent way to define "alternative plans" so that rational choice consists of choosing an optimal plan from a list of alternatives. Local plans cannot be compared meaningfully in terms of their expected utilities.

Second, if we look instead at universal plans, real decision makers would not be able to construct optimal universal plans (even if they exist), both because universal plans are too complex for real agents to construct and because finding optimal ones would require surveying and comparing infinitely many plans.

The upshot is that rational deliberation cannot be expected to produce optimal plans. A decision maker should be an evolutionary planner rather than an optimizing planner. An evolutionary planner finds good plans, and replaces them by better plans as they are found. The concept of a "good plan" and a "better plan" were analyzed in terms of master plans, with the result that the objective of rational deliberation should be to find an acceptable master plan and to be on the continual lookout for ways of improving the master plan.

Real decision makers will not be able to construct master plans as the result of single planning exercises. Master plans are too complex for that. The master plan must instead be constructed incrementally, by engaging in local planning and then merging the local plans into the master plan. The result is the theory of locally global planning.

An important question remains. There can be interactions between local plans. A newly constructed local plan has at least the potential to interact with *all* of my previously adopted plans. Why should constructing a local plan *in the context of a master plan* be any easier than simply constructing a master plan? This will be taken up in chapters 11 and 12, where it will be argued that defeasible principles of causal and probabilistic reasoning already discussed in chapters 5, 7, and 8 enable us to reason defeasibly about master plans, and that is what makes decision-theoretic reasoning computationally feasible for realistically resource-bounded agents.

The main audience for this book is philosophers, but it also bears directly on AI planning theorists, and I would hope that researchers in that area will take it seriously. The preceding remarks establish that existing algorithms for decision-theoretic planning are based on a logically incorrect theory of rational decision making for autonomous agents. As such, they cannot possibly provide the basis for decision-theoretic planning in such agents. What is needed is an evolutionary planner that implements the principle of locally global planning. Existing planning algorithms may work fine for practical applications of limited scope that can be modeled as toy problems, but it is in principle impossible for them to scale up to the kind of planning a sophisticated autonomous agent must do in the real world.

It is important to implement. I doubt that one can ever get a theory of decision-theoretic planning wholly right without testing it by implementing it. But there is such a thing as premature implementation. We need to work out the basics of the theory before we know what to implement. That is what this chapter is about. Doing this work is a necessary precondition to implementing.

I am not yet ready to implement, because there is more theoretical work to be done. However, much of that theoretical work bears directly on how an implementation might go. Chapter 12 takes up the task of sketching in more detail just how locally global planning can actually work. This is both about planning in human beings and planning in artificial agents. As we will see, the problems of making it work are of a computational nature. Humans do make it work, so by investigating how they accomplish that we can get guidance as to how to build an AI planning system that does the same thing.

Chapter 12 investigates locally global planning in more detail, but before going on to that, there is an important loose end to tie up. Locally global planning is formulated in terms of the expected utility of a plan, but that concept has been defined only loosely. Chapter 11 aims to make that more precise. Chapter 11 can be skipped by those not interested in the technical details.

# 11
# Plans and Their Expected Utilities

Cognizers solve practical problems by constructing plans, and the concept of a plan and its expected utility is central to the principle of locally global planning. However, neither concept has been discussed with any precision. Plans were introduced as "courses of action" or "recipes for action", and we defined the expected utility of a plan to be the marginal expected value of trying to execute the plan if one can. The latter cries out for elaboration. What is it to try to execute a plan? We cannot answer that question without first saying more precisely what the structure of a plan is. So this chapter aims to fill these lacunae. However, the details are unavoidably complex, so the reader who does not want to work through the details is invited to go on to chapter 12, which can be understood without reading the present chapter.

## 1. Linear Plans

The simplest plans are *linear plans*, which are finite sequences of temporally ordered actions. The plan given in chapter 10 for making a cup of tea is an example of a linear plan. The plan was: (1) heat water, (2) retrieve a tea bag, (3) place the tea bag in a teapot, (4) pour boiling water into the tea pot, (5) let the concoction sit for several minutes, (6) pour the tea from the teapot into a cup. Let us consider how to define the expected utility of a linear plan.

Where $p$ is a plan, we defined in general:

$$\mathbf{EU}(p) = \mathbf{MEV}(\textit{trying to execute } p \textit{ if the agent can try to execute } p)$$
$$= \mathbf{PROB}(\textit{the agent can try to execute } p) \cdot \mathbf{MEV}(\textit{trying to execute } p \;/\; \textit{the agent can try to execute } p).$$

What is it to try to execute a linear plan? It might be supposed that trying to execute a linear plan consists of trying to execute each action it prescribes, in the right order. Somewhat surprisingly, that doesn't work. Consider the plan for making a cup of tea. In trying to make a cup of tea in accordance with this plan, I may start by putting the water on to boil. Then I open the cupboard to retrieve a tea bag, but discover that I am out of tea. At that point there is nothing I can do to *continue* trying to make a cup of tea. On the other hand, I did try. So trying to execute the plan does not require trying to execute each action it prescribes.

On the other hand, trying to execute a linear plan does seem to require trying to perform the first step. If I cannot even try to heat water, then I cannot try to make a cup of tea. What about the second step? Trying to execute the plan does not require that I try to perform the second step, because I may be unable to try. But suppose I can try to perform the second step. Then does trying to execute the plan require that once I have tried to perform the first step I will go on to try to perform the second step? This depends upon the plan. In the case of the plan for making tea, it is plausible to suppose that if the agent tries to heat water but fails for some reason, then he should not go on to try to perform the next step. In other words, trying to perform the second step only becomes appropriate when the agent knows that the attempt to perform the first step was successful. Such a plan can be regarded as prescribing the second step only on the condition that the first step has been successfully performed. Such plans can be constructed using "contingencies", as described in section 4. However, not all plans will impose such conditions on the execution of their steps. If the first step is intended to achieve a result that cannot be verified by the agent at the time the plan is being executed, then the performance of the second step cannot depend upon knowing that the first step achieved its purpose. For example, the first step might consist of asking a colleague to do something later in the day. If the agent cannot verify that the colleague did as asked, the performance of the next step cannot be dependent on knowing that.

Let us define *simple linear plans* to be linear plans that impose no conditions on their steps, that is, that contain no contingencies. Simple linear plans are executed "blindly". Once the agent has tried to perform the first step, he will try to perform the second step if he can try, and once he has tried to perform the second step he will try to perform the third step if he can try, and so on. In effect, simple linear plans are defined by the following analysis of trying to execute a simple linear plan $\langle A_1,\ldots,A_n\rangle$:

> An agent tries to execute $\langle A_1,\ldots,A_n\rangle$ iff:
> (1) the agent tries to perform $A_1$; and
> (2) for each $i$ such that $1 < i \leq n$, if the agent has tried to perform each $A_j$ for $j < i$, he will subsequently try to perform $A_i$ if he can try.

So when an agent tries to execute a simple linear plan, he will try to perform the first step. Then for each subsequent step, once he has tried to perform all the earlier steps, he will try to perform the next step if he can try. If he cannot try to perform a step, the attempt to execute the plan terminates.

It follows from this that an agent *can* try to execute $\langle A_1,\ldots,A_n\rangle$ iff he *can* try to perform $A_1$. To see this, consider a two-step plan $\langle A_1,A_2\rangle$. The preceding analysis yields:

> An agent tries to execute $\langle A_1,A_2\rangle$ iff:
> (1) the agent tries to perform $A_1$; and
> (2) if the agent has tried to perform $A_1$, he will subsequently try to perform $A_2$ if he can try.

It follows that:

> An agent can try to execute $\langle A_1,A_2\rangle$ iff:
> (1) the agent can try to perform $A_1$; and
> (2) if the agent has tried to perform $A_1$, he will subsequently be able to try to perform $A_2$ if he can try to perform $A_2$.

However, the second clause is a tautology, so it follows that the agent can try to execute $\langle A_1,A_2\rangle$ iff he can try to perform $A_1$. Analogous reasoning establishes the same result for $n$-step plans:

> An agent can try to execute a simple linear plan $\langle A_1,...,A_n\rangle$ iff the agent can try to perform $A_1$.

The most important consequence of this analysis of trying to execute a simple linear plan will be that it enables us to compute the expected utility of the plan in terms of the conditional expected utilities of the actions in the plan. We have seen that forming the intention to perform an action is like adopting the conditional policy of trying to perform the action if one can try, and evaluating the action in terms of its expected utility is the same thing as evaluating the conditional policy in terms of its expected value. Simple linear plans can also be viewed as conditional policies, although of a more complex sort. The next section will investigate "linear policies". It will be shown that the preceding analysis of trying to execute a simple linear plan has the consequence that the expected utility of a linear plan is equal to the marginal expected value of a corresponding "conditional linear policy", and that in turn leads to a characterization of the expected utility of the plan in terms of the expected utilities of its steps.

## 2. Linear Policies

Thus far we have considered how to define the causal probability of an outcome given a single action. In decision-theoretic planning we will often want to consider what is apt to happen if we perform several actions sequentially. Let *linear policies* be sequences of actions in which each action postdates its predecessor. First consider a linear policy consisting of two actions $A_1,A_2$. Computing the causal probability of an outcome is complicated by the fact that some constituents of the background of the second action may postdate the first action, and the first action can affect the probabilities of those constituents. So let $\mathbf{B}$ be the set of backgrounds for $A_1$ conjoined with those parts of the backgrounds of $A_2$ that predate $A_1$, and let $\mathbf{B^*}$ consist of the remainders of the backgrounds for $A_2$. The members of $\mathbf{B^*}$ postdate $A_1$. On analogy to figure 8.1 of chapter 8, if $O$ postdates $A_2$ then we can employ figure 11.1 to conceptualize the world as unfolding temporally. A *scenario* is a path through the tree. The probability associated with a scenario should be

$$\textbf{PROB}(B_i)\cdot\textbf{PROB}(B^*_j / A_1\&B_i)\cdot\textbf{PROB}(O / A_1\&A_2\&B_i\&B^*_j).$$

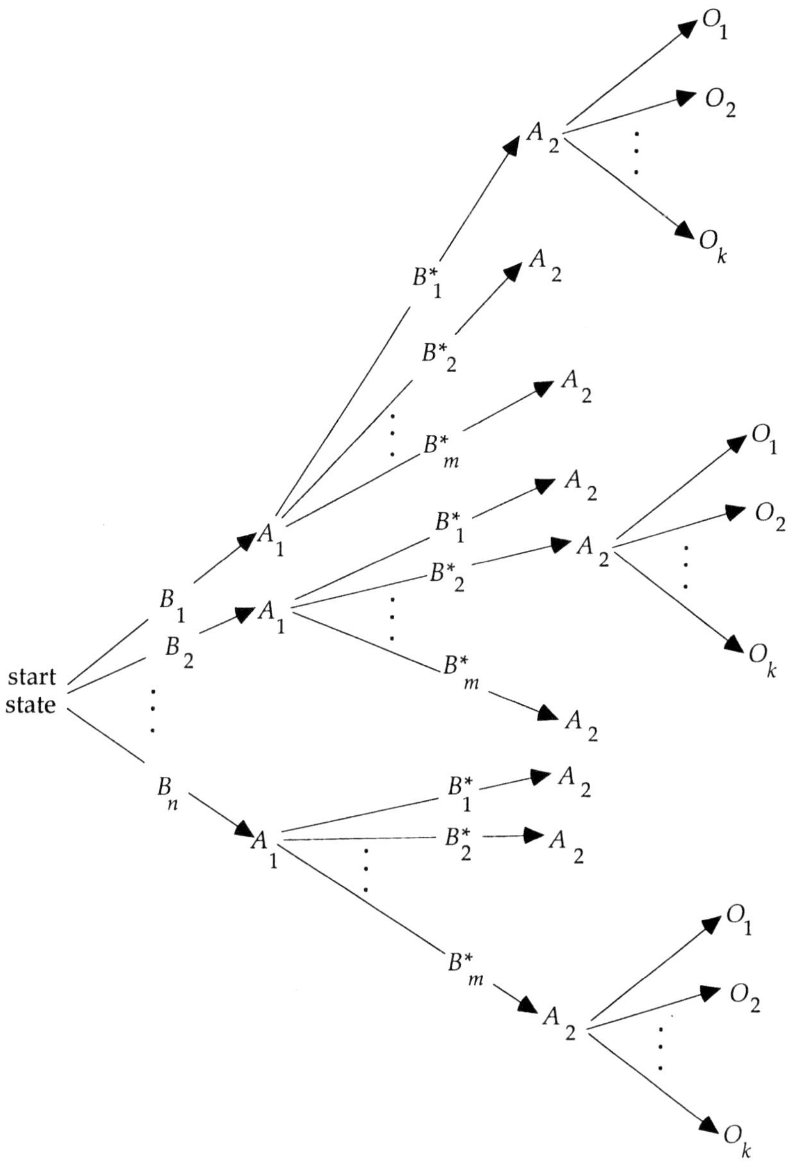


Figure 11.1. Scenarios with two actions

Then we can define:

$$\textsf{C-PROB}_{A_1,A_2}(O)$$
$$= \sum_{B\in \mathbf{B}} \textsf{PROB}(B) \cdot \sum_{B^*\in \mathbf{B}^*} \textsf{PROB}(B^*/A_1\&B)\cdot\textsf{PROB}(O/A_1\&A_2\&B\&B^*)$$
$$= \sum_{B\in \mathbf{B}} \textsf{PROB}(B)\cdot\textsf{C-PROB}_{A_2}(O/A_1\&B).$$

This definition can be generalized recursively to arbitrary sequences (for $k > 1$) of actions postdated by $O$:

$$\textsf{C-PROB}_{A_1,\ldots,A_k}(O) = \sum_{B\in \mathbf{B}} \textsf{PROB}(B)\cdot\textsf{C-PROB}_{A_2,\ldots,A_k}(O/A_1\&B).$$
$$\textsf{C-PROB}_{A_1,\ldots,A_k}(O/Q) = \sum_{B\in \mathbf{B}} \textsf{PROB}(B/Q)\cdot\textsf{C-PROB}_{A_2,\ldots,A_k}(O/A_1\&B\&Q).$$

This calculation is what we get from propagating probabilities through scenarios in temporal order.

As thus-far construed, linear policies are simple sequences of actions. However, simple linear plans have a more complex structure. They can be viewed as *conditional linear policies*, which are sequences of conditional policies rather than sequences of actions. Let $A_1$ *if* $C_1,\ldots,A_k$ *if* $C_k$ be the policy *do* $A_1$ *if* $C_1$, *then do* $A_2$ *if* $C_2$, *then* … . As for simple conditional policies (see chapter 8), the probabilities of the $B_i$'s and the probabilities of the outcomes must be made conditional on the $C_i$'s. If $C_i$ is false, the rest of the policy will still be executed. So on analogy to simple conditional policies, for $k > 1$ the causal probability can be defined recursively as:

$$\textsf{C-PROB}_{A_1\, if\, C_1,\ldots,A_k\, if\, C_k}(O)$$
$$= \left[\textsf{PROB}(C_1)\cdot\sum_{B\in \mathbf{B}} \textsf{PROB}(B/C_1)\cdot\textsf{C-PROB}_{A_2\, if\, C_2,\ldots,A_k\, if\, C_k}(O/A_1\&B\&C_1)\right]$$
$$+ \left[\textsf{PROB}(\sim C_1)\cdot\textsf{C-PROB}_{A_2\, if\, C_2,\ldots,A_k\, if\, C_k}(O/\sim C_1)\right].$$

$$\textsf{C-PROB}_{A_1\, if\, C_1,\ldots,A_k\, if\, C_k}(O/Q)$$
$$= \left[\textsf{PROB}(C_1/Q)\right.$$
$$\left.\cdot\sum_{B\in \mathbf{B}} \textsf{PROB}(B/C_1\&Q)\cdot\textsf{C-PROB}_{A_2\, if\, C_2,\ldots,A_k\, if\, C_k}(O/A_1\&B\&C_1\&Q)\right]$$
$$+ \left[\textsf{PROB}(\sim C_1/Q)\cdot\textsf{C-PROB}_{A_2\, if\, C_2,\ldots,A_k\, if\, C_k}(O/\sim C_1\&Q)\right].$$

We can then define the expected value of a conditional linear policy in terms of these causal probabilities:

$$\mathbf{EV}(A_1 \textit{ if } C_1,\ldots,A_k \textit{ if } C_k) = \sum_{O\in\mathbf{O}} \mathrm{U}(O)\cdot\textbf{C-PROB}_{A_1 \textit{ if } C_1,\ldots,A_k \textit{ if } C_k}(O).$$

Now let us apply this to simple linear plans. The expected utility of a simple linear plan $\langle A_1,\ldots,A_n\rangle$ is **MEV**(*the agent tries to execute* $\langle A_1,\ldots,A_n\rangle$ *if he can try to execute* $\langle A_1,\ldots,A_n\rangle$). We have seen that the agent can try to execute $\langle A_1,\ldots,A_n\rangle$ iff he can try to execute $A_1$. It was argued in section 1 that:

> An agent tries to execute $\langle A_1,\ldots,A_n\rangle$ iff:
> (1) the agent tries to perform $A_1$; and
> (2) for each $i$ such that $1 < i \leq n$, if the agent has tried to perform each $A_j$ for $j < i$, he will subsequently try to perform $A_i$ if he can try.

It follows that:

> An agent tries to execute $\langle A_1,\ldots,A_n\rangle$ if he can try to perform $A_1$ iff:
> (1) the agent tries to perform $A_1$ if he can try to perform $A_1$; and
> (2) for each $i$ such that $1 < i \leq n$, if the agent has tried to perform each $A_j$ for $j < i$, he will subsequently try to perform $A_i$ if he can try.

Equivalently:

> An agent tries to execute $\langle A_1,\ldots,A_n\rangle$ if he can try to execute $A_1$ iff for each $i$ such that $1 \leq i \leq n$, if the agent has tried to perform each $A_j$ for $j < i$, he will subsequently try to perform $A_i$ if he can try.

Thus adopting the plan is equivalent to adopting the conditional linear policy *try-*$A_1$ *if can-try-*$A_1$, *try-*$A_2$ *if (can-try-*$A_2$ & *try-*$A_1$), ... , *try-*$A_n$ *if (can-try-*$A_n$ & *try-*$A_1$ & ... & *try-*$A_{n-1}$). Precisely:

**Theorem 1:** If $\langle A_1,\ldots,A_n\rangle$ is a simple linear plan then

$$\begin{aligned}&\mathbf{EU}(\langle A_1,\ldots,A_n\rangle)\\ &= \mathbf{MEV}(\textit{try-}A_1 \textit{ if can-try-}A_1, \ldots,\\ &\qquad \textit{try-}A_n \textit{ if (can-try-}A_n \ \& \ \textit{try-}A_1 \ \& \ldots \& \ \textit{try-}A_{n-})).\end{aligned}$$

We saw in chapter 8 that deciding to perform an action amounts to deciding to try to perform it provided one can try to perform it. Because of the failure of action omnipotence, that is the most we can be sure of doing. Theorem 1 tells us that we can think of simple linear plans analogously. That is, adopting the plan amounts to deciding to try to perform its constituent actions at the appropriate times provided we can try to do so.

The point of representing plans as conditional linear policies is that we have a precise definition of the expected value of such a policy. We can appeal to that definition to prove theorems about the expected utility of a plan and to compute its value.

## 3. Nonlinear Plans

Section 2 focused on "simple linear" plans, where if the agent is unable to try to perform a step then he stops trying to execute the rest of the plan, but otherwise he continues. However, most plans have more complex structures. First, plans need not be linear. Some steps of the plan can be unordered with respect to others. Using an example from Russell and Norvig (1995), a plan to put on your shoes and socks might look as in figure 11.2. Here *put on right sock* is ordered before *put on right shoe*, but it is unordered with respect to either *put on left sock* or *put on left shoe*. When the plan is actually executed, the steps will be performed in some order, but the decision of which steps to perform first can be left until the time of execution. Such plans are *nonlinear plans*.

Nonlinear plans are generally constructed by merging smaller linear plans into a single plan. For example, the above plan might be produced by merging a plan to put on my right shoe and sock and a plan to put on my left shoe and sock. When plans are merged, the result will normally be a nonlinear plan because steps drawn from different constituent subplans will usually be unordered with respect to each other.

A *linearization* of a nonlinear plan is a linear plan that results from adding ordering constraints to the nonlinear plan to linearly order the plan steps. The expected utility of a nonlinear plan ought to be determined by the expected utilities of its linearizations. If a planning agent discovers that some linearizations have lower expected utilities than others, then he should add ordering constraints that preclude those linearizations. So it is reasonable to define the expected utility of a nonlinear plan to be the maximum of the expected utilities of its linearizations. To the best of the agent's knowledge, all the linearizations should have the same expected utility.

The linearizations of a nonlinear plan will not normally be simple linear plans. If we are unable to try to perform a step of one of the constituent plans out of which the nonlinear plan is constructed, that should abort execution of the rest of that constituent plan, but may not abort execution of the other constituent plans. For example, if I cannot find my right sock, I may still try to don my left sock. What this illustrates is that in general, plans will incorporate a dependency relation on steps, and if the agent is unable to try to perform a step, he will only try to execute the remaining steps that do not depend upon the failed step. A step can have several purposes in a plan, and for each purpose it may depend upon a different

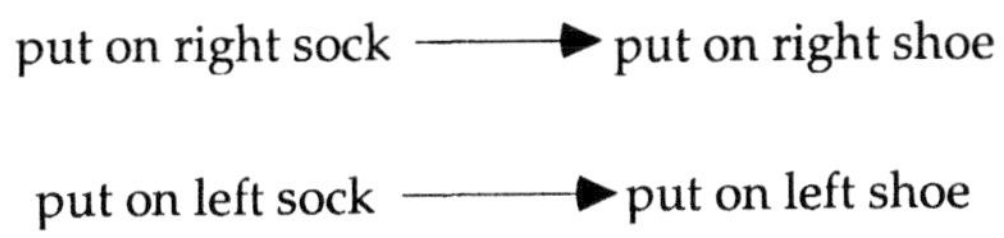


Figure 11.2. A nonlinear plan

set of steps. We can handle this by talking about *dependency-sets* for plan-steps. A step can have multiple dependency sets corresponding to multiple purposes, and even if some of the purposes are thwarted the plan may still dictate performing the step in order to achieve the remaining purposes. The *initial steps* of a plan will have empty dependency-sets and no steps will be ordered before them. Let us define recursively that a step is *cancelled* iff either the agent was unable to try to perform it or every dependency-set for it contains some cancelled step. Then we might try defining:

> An agent tries to execute a plan iff for each step $S$ of the plan, if the agent has tried to perform all of the steps in some dependency-set for $S$ and all the uncancelled steps ordered before $S$ in the plan, then he tries to perform $S$ if he can try.

This has the consequence that in order to try to execute a plan, the agent must try to execute all of the initial steps. However, this definition does not quite capture what we want. For example, if we apply this to simple linear plans, each step $S_{i+1}$ will have a single dependency-set containing only its immediate predecessor $S_i$. The preceding analysis of trying to execute a plan would then make it equivalent to executing a linear policy whose clauses have the form *try*-$A_n$ *if can-try*-$A_n$ & *try*-$A_{n-1}$. However, as we have seen, the clauses of the linear policy ought to have the form *try*-$A_n$ *if can-try*-$A_n$ & *try*-$A_1$ & ... & *try*-$A_{n-1}$. This will make an important difference to the policy. The policy whose clauses have the form *try*-$A_n$ *if can-try*-$A_n$ & *try*-$A_{n-1}$ would have us consider the possibility that although the agent was unable to try to perform some of the earlier steps, and so the later steps were not "called" by the plan, he nevertheless performs *try*-$A_{n-1}$ for some other reason. The policy would then require him to go on and perform *try*-$A_n$ if he can. This is not required by the policy whose clauses have the form *try*-$A_n$ *if can-try*-$A_n$ & *try*-$A_1$ & ... & *try*-$A_{n-1}$. Only the latter provides the correct analysis of what it is to try to execute the simple linear policy.

We can express the preceding observations by saying that the policy should only require the agent to try to perform a step if it is "called" by the policy. In the simple linear policy, *try*-$A_{n-1}$ is called just in case the agent tried to perform all the previous steps, but in nonlinear plans the calling relation is more complex. In general, it can be defined recursively as follows:

> A step $S$ is *called* iff for some dependency-set, (1) every step in the dependency-set is called and the agent tried to perform it, and (2) for every step ordered before $S$ in the plan, either it is cancelled or it has been called and the agent tried to perform it if he could try.

Then we can repair the flawed analysis of trying to execute a plan as follows:

> An agent tries to execute a plan iff for each step $S$ of the plan, if $S$ is called then the agent tries to perform $S$ if he can.

A further refinement will be required when, in the next section, we consider conditional plans.

# 4. Conditional Plans

To produce plans with reasonable expected utilities, it will be important to add a further complication to the logical structure of plans. In planning what to do, we often lack knowledge that would help us make a better decision. For example, in planning a route for driving from one point to another, I may be unsure whether a certain road is under construction. Ideally, I should plan to go one way if the road is under construction, but a different way if it is not. We handle this by planning for both possibilities, and then decide at the time we execute the plan which way to go. This can be accommodated by inserting *contingencies* into a plan.[1] A contingency is an epistemic condition—normally that the agent does or does not believe some proposition *P*—and then some of the subsequent plan steps are made dependent on it. At execution time, those plan steps are only executed if the agent satisfies the epistemic condition. Let us say that the agent *satisfies* a contingency if he satisfies the condition labeling it.

To accommodate the dependency of plan-steps on contingencies, we can include contingencies in the dependency-sets. We expand the definition of cancellation by saying that a step is *cancelled* iff either the agent was unable to try to perform it or every dependency-set for it contains some cancelled step or some contingency the agent does not satisfy. Then we can revise the definition of a step being called as follows:

> A step *S* is *called* iff (1) for some dependency-set, all the contingencies are satisfied, and every step in the dependency-set is called and the agent tried to perform it, and (2) for every step ordered before *S* in the plan, either it is cancelled or it has been called and the agent tried to perform it if he could try.

The preceding definition has the consequence that initial steps may not be called, because they may depend upon contingencies that are not satisfied. If none of them are called, then even if the agent tries to perform every step of the plan that is called, he may do nothing. If he does nothing, he has not tried to execute the plan. So we must modify the definition of trying to execute a plan as follows:

> An agent tries to execute a plan iff (1) some initial step is called, and (2) for each step *S* of the plan, if *S* is called then the agent tries to perform *S* if he can.

Corresponding to each step *S* of a (possibly nonlinear) plan is the conditional policy *try to perform S if* (1) *you can try,* (2) *you satisfy all the contingencies in some dependency-set for S,* (3) *all the steps in the dependency-set have been*

---

[1] This basic idea derives originally from work on classical contingency planning by Warren (1976), Peot and Smith (1992), and Pryor and Collins (1996). Attempts to combine contingency planning with decision-theoretic planning have been made by Draper, Hanks, and Weld (1994), Blythe and Veloso (1997), and Onder and Pollack (1997, 1999).

*called and you have already tried to perform them, and* (4) *all the uncancelled steps ordered before S in the plan have been called and you have tried to perform them.* However, collecting these together does not generate a conditional linear policy because the steps need not be linearly ordered. If a plan is nonlinear then there will be points in the course of its execution at which the agent has a choice of what to do next. These will be cases in which the conditions of more than one of these conditional policies are satisfied simultaneously. Accordingly, we cannot apply results pertaining to conditional linear policies directly to the evaluation of the plan.

This problem can be circumvented by looking at the linearizations of the plan. A linear plan corresponds to the conditional linear policy that is the sequence of the conditional policies corresponding to the steps of the plan, and the expected utility of the linearization is the marginal expected value of this policy. This enables us to use general theorems about conditional linear policies to characterize the expected utility of a plan.

Recall that causal probability was defined for conditional linear policies as follows:

$$\textbf{C-PROB}_{A_1 \text{ if } C_1,\ldots,A_k \text{ if } C_k}(O)$$
$$= \left[\textbf{PROB}(C_1)\cdot\sum_{B\in\mathbf{B}} \textbf{PROB}(B/C_1)\cdot\textbf{C-PROB}_{A_2 \text{ if } C_2,\ldots,A_k \text{ if } C_k}(O/A_1\&B\&C_1)\right]$$
$$+ \left[\textbf{PROB}(\sim C_1)\cdot\textbf{C-PROB}_{A_2 \text{ if } C_2,\ldots,A_k \text{ if } C_k}(O/\sim C_1).\right.$$

$$\textbf{C-PROB}_{A_1 \text{ if } C_1,\ldots,A_k \text{ if } C_k}(O/Q)]$$
$$= \left[\textbf{PROB}(C_1/Q)\right.$$
$$\left.\cdot\sum_{B\in\mathbf{B}} \textbf{PROB}(B/C_1\&Q)\cdot\textbf{C-PROB}_{A_2 \text{ if } C_2,\ldots,A_k \text{ if } C_k}(O/A_1\&B\&C_1\&Q)\right]$$
$$+ \left[\textbf{PROB}(\sim C_1/Q)\cdot\textbf{C-PROB}_{A_2 \text{ if } C_2,\ldots,A_k \text{ if } C_k}(O/\sim C_1\&Q)\right].$$

For simple linear policies, scenarios as diagrammed in figure one above have the form

$$B_1 \to A_1 \to B_2 \to A_2 \to \ldots \to B_k \to A_k.$$

Scenarios for conditional linear policies are more complex, because they must take account of whether the conditions are satisfied. Each condition may be either satisfied or unsatisfied, and if it is unsatisfied then the corresponding action is not included in the scenario. This means that a scenario for a conditional linear policy looks like this:

$$(\sim)C_1 \to B_1 \to (A_1) \to (\sim)C_2 \to B_2 \to (A_2) \to \ldots \to (\sim)C_k \to B_k \to (A_k)$$

where each tilde can be present or absent, and $A_i$ is included in the scenario iff the tilde is absent on $C_i$. A scenario is characterized by the set of unnegated

$C_i$'s. For example, the following is a scenario:

$$C_1 \to B_1 \to A_1 \to \sim C_2 \to B_2 \to C_3 \to B_3 \to A_3$$

Given a scenario $S$, let $C_S$ be the conjunction of the $C_i$'s, $\sim C_i$'s, and $B_i$'s in the scenario, and let $A_S$ be the conjunction of the actions in the scenario. Defining the probability of a scenario as $\textbf{C-PROB}_{A_1 \textit{ if } C_1,\ldots,A_k \textit{ if } C_k}(C_S)$, we get:

**Theorem 2:** If **SC** is the set of scenarios for the policy $A_1$ *if* $C_1,\ldots,A_k$ *if* $C_k$ relative to $O$ then

$$\textbf{C-PROB}_{A_1 \textit{ if } C_1,\ldots,A_k \textit{ if } C_k}(O/Q)$$
$$= \sum_{S\in \mathbf{SC}} \textbf{C-PROB}_{A_1 \textit{ if } C_1,\ldots,A_k \textit{ if } C_k}(S/Q)\cdot\textbf{PROB}(O/C_S \& A_S \& Q).$$

It then follows that:

**Theorem 3:** If **SC** is the set of scenarios for the policy $A_1$ *if* $C_1,\ldots,A_k$ *if* $C_k$ then:

$$\textbf{MEV}(A_1 \textit{ if } C_1,\ldots,A_k \textit{ if } C_k) = \sum_{S\in \mathbf{SC}} \textbf{C-PROB}_{A_1 \textit{ if } C_1,\ldots,A_k \textit{ if } C_k}(S)\cdot\textbf{MEV}(S).$$

Thus the marginal expected value of a conditional linear policy is a weighted average of the marginal expected values of its scenarios.

The next theorem tells us that the marginal expected value of a scenario is the sum of the marginal expected values of its actions in the context of the scenario:

**Theorem 4:** If $S$ is a scenario and $A_1,\ldots,A_n$ are the actions it prescribes listed in temporal order then:

$$\textbf{MEV}(S) = \sum_{1\le i\le n} \textbf{MEV}(A_i/A_1 \& \ldots \& A_{i-1} \& C_S).$$

This theorem provides the basic mechanism for computing the marginal expected value of a linear policy. We can use it to compute the expected utility of a plan. In the next section, I will illustrate this with an example.

## 5. An Example

Theorems 3 and 4 together tell us how to compute the marginal expected value of a linear policy, and hence how to compute the expected utility of a plan, in terms of the marginal expected values of the actions constituting the policy in the possible scenarios of the policy. Unfortunately, the computation this prescribes will often be very difficult. It requires us to compute marginal expected values for every scenario separately, and compute the marginal expected value of the policy or plan as a weighted average of the marginal expected values of the scenarios. There can be a very large number of scenarios. Thus efficient decision-theoretic planning requires more efficient ways of computing expected utilities. My proposal is that direct inference provides a mechanism for doing this defeasibly. Rather than work out a general account of this, I will illustrate it with an example.

Consider a simple planning problem. The agent is on a target range,with a target pistol, and the goal is to hit the target. To do this the agent must load the gun, aim the gun, and pull the trigger. A complication arises in that there is a warning light over the target the glows red when it is not safe to shoot, and turns green when it is safe to shoot. The agent may construct the nonlinear plan diagrammed in figure 11.3. The plan aims at achieving certain subgoals in order to make it probable that subsequent actions can be performed, and tries to achieve other subgoals in order to make it probable that actions will have their desired effects. The heavy dashed arrow between *try to load gun* and *look at light* signifies an ordering constraint—the latter is

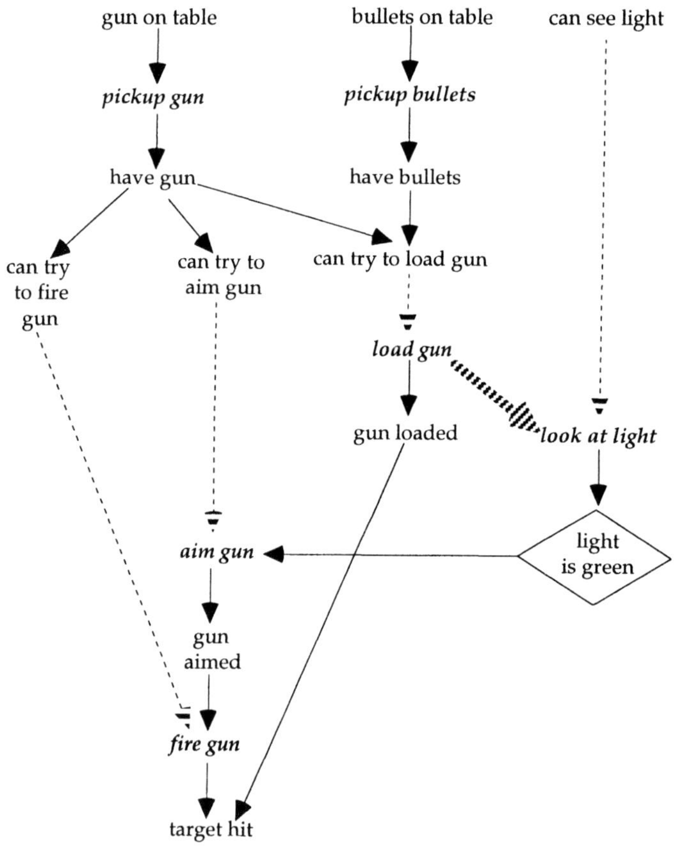


Figure 11.3. Plan for hitting target

not to be done until the former has been done. The arrow from the diamond containing *light is green* to *try to aim gun* signifies that the latter is not to be done until the agent believes the light is green.

This is a nonlinear plan because *pickup gun* and *pickup bullets* are not ordered with respect to each other. To convert this into a conditional linear policy, we must arbitrarily order *pickup gun* and *pickup bullets* with respect to each other. If we order the former before the latter, we obtain a linear plan. The corresponding conditional linear policy is then:

*try-to-pickup-gun if can-try-to-pickup-gun*
*try-to-pickup-bullets*
  *if* [(*can-try-to-pickup-bullets* & ~*can-try-to-pickup-gun*)
   ∨ (*can-try-to-pickup-bullets* & *try-to-pickup-gun*)]
*try-to-load-gun*
  *if* (*can-try-to-load-gun* & *try-to-pickup-gun* & *try-to-pickup-bullets*)
*try-to-look-at-light*
  *if* [(*can-try-to-look-at-light* & *try-to-pickup-gun*
    & *try-to-pickup-bullets* & ~*can-try-to-load-gun*)
   ∨ (*can-try-to-look-at-light* & *try-to-pickup-gun*
    & *try-to-pickup-bullets* & *try-to-load-gun*)]
*try-to-aim-gun*
  *if* [(*can-try-to-aim-gun* & *try-to-pickup-gun* & *try-to-pickup-bullets*
    & *try-to-load-gun* & *believe-light-is-green*
    & ~*can-try-to-look-at-light*)
   ∨ (*can-try-to-aim-gun* & *try-to-pickup-gun* & *try-to-pickup-bullets*
    & *try-to-load-gun* & *believe-light-is-green* & *try-to-look-at-light*)]
*try-to-fire-gun*
  *if* [(*can-try-to-fire-gun* & *try-to-pickup-gun* & *try-to-pickup-bullets*
    & *try-to-load-gun* & *believe-light-is-green*
    & ~*can-try-to-look-at-light* & *try-to-aim-gun*)
   ∨ (*can-try-to-fire-gun* & *try-to-pickup-gun* & *try-to-pickup-bullets*
    & *try-to-load-gun* & *believe-light-is-green* & *try-to-look-at-light*
    & *try-to-aim-gun*)]

The causal probability, relative to this policy, of the target being hit expands into a long sum of the following form:

$$\Big\{\sum_{B_1 \in \mathbf{B}_1} \mathbf{PROB}(B_1 \;\&\; \textit{can-try-to-pickup-gun})$$

$$\cdot \Big[\Big(\sum_{B_2 \in \mathbf{B}_2} \mathbf{PROB}(B_2 \;\&\; \textit{can-try-to-pickup-bullets} \;/\; \textit{try-to-pickup-gun} \;\&\; B_1)$$

$$\cdot \Big\{\Big[\sum_{B_3 \in \mathbf{B}_3} \mathbf{PROB}(B_3 \;\&\; \textit{can-try-to-load-gun} \;/\; \textit{try-to-pickup-gun} \;\&\; B_1$$

$$\& \mathit{try\text{-}to\text{-}pickup\text{-}bullets} \ \& \ B_2)$$

$$\cdot \Big(\Big\{\sum_{B_4 \in \mathbf{B}_4} \mathbf{PROB}(B_4 \ \& \ \mathit{can\text{-}try\text{-}to\text{-}look\text{-}at\text{-}light} \ /$$

$$\mathit{try\text{-}to\text{-}pickup\text{-}gun} \ \& \ B_1 \ \& \ \mathit{try\text{-}to\text{-}pickup\text{-}bullets} \ \& \ B_2 \ \& \ \mathit{try\text{-}to\text{-}load\text{-}gun} \ \& \ B_3)$$

$$\cdot \Big[\Big(\sum_{B_5 \in \mathbf{B}_5} \mathbf{PROB}(B_5 \ \& \ \mathit{can\text{-}try\text{-}to\text{-}aim\text{-}gun} \ \& \ \mathit{believe\text{-}light\text{-}is\text{-}green} \ /$$

$$\mathit{try\text{-}to\text{-}pickup\text{-}gun} \ \& \ B_1 \ \& \ \mathit{try\text{-}to\text{-}pickup\text{-}bullets} \ \& \ B_2 \ \& \ \mathit{try\text{-}to\text{-}load\text{-}gun} \ \& \ B_3 \ \& \ \mathit{try\text{-}to\text{-}look\text{-}at\text{-}light} \ \& \ B_4)$$

$$\cdot \Big\{\Big[\sum_{B_6 \in \mathbf{B}_6} \mathbf{PROB}(B_6 \ \& \ \mathit{can\text{-}try\text{-}to\text{-}fire\text{-}gun} \ / \ \mathit{try\text{-}to\text{-}pickup\text{-}gun} \ \& \ B_1 \ \& \ \mathit{try\text{-}to\text{-}pickup\text{-}bullets} \ \& \ B_2 \ \& \ \mathit{try\text{-}to\text{-}load\text{-}gun} \ \& \ B_3 \ \& \ \mathit{try\text{-}to\text{-}look\text{-}at\text{-}light} \ \& \ B_4 \ \& \ \mathit{believe\text{-}light\text{-}is\text{-}green} \ \& \ \mathit{try\text{-}to\text{-}aim\text{-}gun} \ \& \ B_5)$$

$$\cdot \mathbf{PROB}(\mathit{target\text{-}hit} \ / \ \mathit{try\text{-}to\text{-}pickup\text{-}gun} \ \& \ B_1 \ \& \ \mathit{try\text{-}to\text{-}pickup\text{-}bullets} \ \& \ B_2 \ \& \ \mathit{try\text{-}to\text{-}load\text{-}gun} \ \& \ B_3 \ \& \ \mathit{try\text{-}to\text{-}look\text{-}at\text{-}light} \ \& \ B_4 \ \& \ \mathit{believe\text{-}light\text{-}is\text{-}green} \ \& \ \mathit{try\text{-}to\text{-}aim\text{-}gun} \ \& \ B_5 \ \& \ \mathit{try\text{-}to\text{-}fire\text{-}gun} \ \& \ B_6)\Big]$$

...

$$+ \Big\{\sum_{B_1 \in \mathbf{B}_1} \mathbf{PROB}(B_1 \ \& \ {\sim}\mathit{can\text{-}try\text{-}to\text{-}pickup\text{-}gun})$$

$$\cdot \Big\{\Big[\sum_{B_2 \in \mathbf{B}_2} \mathbf{PROB}(B_2 \ \& \ \mathit{can\text{-}try\text{-}to\text{-}pickup\text{-}bullets} \ / \ {\sim}\mathit{can\text{-}try\text{-}to\text{-}pickup\text{-}gun} \ \& \ B_1)$$

$$\cdot \sum_{B_3 \in \mathbf{B}_3} \mathbf{PROB}(B_3 \ / \ {\sim}\mathit{can\text{-}try\text{-}to\text{-}pickup\text{-}gun} \ \& \ B_1 \ \& \ \mathit{try\text{-}to\text{-}pickup\text{-}bullets} \ \& \ B_2)$$

$$\cdot \sum_{B_4 \in \mathbf{B}_4} \mathbf{PROB}(B_4 \ / \ {\sim}\mathit{can\text{-}try\text{-}to\text{-}pickup\text{-}gun} \ \& \ B_1 \ \& \ \mathit{try\text{-}to\text{-}pickup\text{-}bullets} \ \& \ B_2 \ \& \ B_3)$$

$$\cdot \sum_{B_5 \in \mathbf{B}_5} \mathbf{PROB}(B_5 \ / \ {\sim}\mathit{can\text{-}try\text{-}to\text{-}pickup\text{-}gun} \ \& \ B_1 \ \& \ \mathit{try\text{-}to\text{-}pickup\text{-}bullets} \ \& \ B_2 \ \& \ B_3 \ \& \ B_4)$$

$$\cdot \sum_{B_6 \in \mathbf{B}_6} \mathbf{PROB}(B_6 \ / \ {\sim}\mathit{can\text{-}try\text{-}to\text{-}pickup\text{-}gun} \ \& \ B_1 \ \& \ \mathit{try\text{-}to\text{-}pickup\text{-}bullets} \ \& \ B_2 \ \& \ B_3 \ \& \ B_4 \ \& \ B_5)$$

$$\cdot \mathbf{PROB}(\mathit{target\text{-}hit} \ / \ {\sim}\mathit{can\text{-}try\text{-}to\text{-}pickup\text{-}gun} \ \& \ B_1 \ \& \ \mathit{try\text{-}to\text{-}pickup\text{-}bullets} \ \& \ B_2 \ \& \ B_3 \ \& \ B_4 \ \& \ B_5 \ \& \ B_6)\Big]$$

$$+ \Big[\sum_{B_2 \in \mathbf{B}_2} \mathbf{PROB}(B_2 \ \& \ {\sim}\mathit{can\text{-}try\text{-}to\text{-}pickup\text{-}bullets} \ / \ {\sim}\mathit{can\text{-}try\text{-}to\text{-}pickup\text{-}gun} \ \& \ B_1)$$

$$\cdot \sum_{B_3 \in \mathbf{B}_3} \mathbf{PROB}(B_3 \ / \ {\sim}\mathit{can\text{-}try\text{-}to\text{-}pickup\text{-}gun} \ \& \ B_1 \ \& \ {\sim}\mathit{can\text{-}try\text{-}to\text{-}pickup\text{-}bullets} \ \& \ B_2)$$

$$\cdot \sum_{B_4 \in \mathbf{B}_4} \mathbf{PROB}(B_4 \ / \ {\sim}\mathit{can\text{-}try\text{-}to\text{-}pickup\text{-}gun} \ \& \ B_1 \ \& \ {\sim}\mathit{can\text{-}try\text{-}to\text{-}pickup\text{-}bullets} \ \& \ B_2 \ \& \ B_3)$$

$$\cdot \sum_{B_5 \in \mathbf{B}_5} \mathbf{PROB}(B_5 \ / \ {\sim}\mathit{can\text{-}try\text{-}to\text{-}pickup\text{-}gun} \ \& \ B_1 \ \& \ {\sim}\mathit{can\text{-}try\text{-}to\text{-}pickup\text{-}bullets}$$

$$\& B_2 \& B_3 \& B_4)$$

$$\cdot \sum_{B_6 \in \mathbf{B}_6} \mathbf{PROB}(B_6 \,/\, \sim\!\textit{can-try-to-pickup-gun} \,\&\, B_1 \,\&\, \sim\!\textit{can-try-to-pickup-bullets} \,\&\, B_2 \,\&\, B_3 \,\&\, B_4 \,\&\, B_5)$$

$$\cdot \mathbf{PROB}(\textit{target-hit} \,/\, \sim\!\textit{can-try-to-pickup-gun} \,\&\, B_1 \,\&\, \sim\!\textit{can-try-to-pickup-bullets} \,\&\, B_2 \,\&\, B_3 \,\&\, B_4 \,\&\, B_5 \,\&\, B_6)]]\Big\}$$

To compute the value of this sum, we must be able to evaluate the probabilities that occur in it. The backgrounds $B_1$, $B_2$, $B_3$, $B_4$, $B_5$ and $B_6$ include everything that is relevant to the various actions achieving the specified results. For example, $B_6$ includes *have-gun, gun-loaded* and *gun-aimed*. We would normally suppose that is all that is relevant to whether *try-to-fire-gun* will achieve *target-hit*. That is, by direct inference we would infer:

$$\mathbf{PROB}(\textit{target-hit} \,/\, \textit{try-to-pickup-gun} \,\&\, B_1 \,\&\, \textit{try-to-pickup-bullets} \,\&\, B_2 \,\&\, \textit{try-to-load-gun} \,\&\, B_3 \,\&\, \textit{try-to-look-at-light} \,\&\, B_4 \,\&\, \textit{believe-light-is-green} \,\&\, \textit{try-to-aim-gun} \,\&\, B_5 \,\&\, \sim\!\textit{can-try-to-fire-gun} \,\&\, B_6)$$
$$= \mathbf{PROB}(\textit{target-hit} \,/\, \textit{try-to-fire-gun} \,\&\, \textit{have-gun} \,\&\, \textit{gun-aimed} \,\&\, \textit{gun-loaded}).$$

Direct inference makes it defeasibly reasonable to replace the reference to the backgrounds in the sum by reference to just those factors we know to be relevant. Suppose the only relevant probabilities we know are:

**PROB**(*have-gun* / (*try-to-pickup-gun* & *gun-on-table*)) = 0.9
**PROB**(*have-bullets* / (*try-to-pickup-bullets* & *bullets-on-table*)) = 0.9
**PROB**(*gun-aimed* / *try-to-aim-gun*) = 0.99
**PROB**(*gun-loaded* / *try-to-load-gun*) = 0.98
**PROB**(*target-hit* / (*try-to-fire-gun* & *gun-aimed* & *gun-loaded*)) = 0.8
**PROB**(*believe-light-is-green* /
(*try-to-look-at-light* & *can-see-light* & *light-green*)) = 0.98
**PROB**(*believe-light-is-green* /
(*try-to-look-at-light* & *can-see-light* & ~*light-green*)) = 0.02

Then we can conclude defeasibly that the backgrounds for the actions involved in the plan-steps are as follows:

*try-to-look-at-light:*
{~*light-green can-see-light*}
{*light-green can-see-light*}
{*light-green ~can-see-light*}
{~*light-green ~can-see-light*}
*try-to-fire-gun:*
{*have-gun gun-aimed gun-loaded*}
{*have-gun gun-aimed ~gun-loaded*}

*{have-gun ~gun-loaded ~gun-aimed}*
*{have-gun gun-loaded ~gun-aimed}*
*{gun-loaded ~gun-aimed ~have-gun}*
*{~gun-loaded ~gun-aimed ~have-gun}*
*{gun-aimed ~gun-loaded ~have-gun}*
*{gun-aimed gun-loaded ~have-gun}*

*try-to-load-gun:*
*{have-bullets have-gun}*
*{have-bullets ~have-gun}*
*{~have-gun ~have-bullets}*
*{have-gun ~have-bullets}*

*try-to-aim-gun:*
*{have-gun}*
*{~have-gun}*

*try-to-pickup-bullets:*
*{bullets-on-table}*
*{~bullets-on-table}*

*try-to-pickup-gun:*
*{gun-on-table}*
*{~gun-on-table}*

This enables us to replace reference to the backgrounds by these specific lists of propositions, expanding the sum into a very long sum a small part of which is as follows:

{**PROB**(*gun-on-table* & *can-try-to-pickup-gun*)

· [(**PROB**(*bullets-on-table* & *can-try-to-pickup-bullets* / *try-to-pickup-gun* & *gun-on-table*)

· {[**PROB**(*have-bullets* & *have-gun* & *can-try-to-load-gun* / *try-to-pickup-gun* & *gun-on-table* & *try-to-pickup-bullets* & *bullets-on-table*)

· ({**PROB**(*can-see-light* & *light-green* & *can-try-to-look-at-light* / *try-to-pickup-gun* & *gun-on-table* & *try-to-pickup-bullets* & *bullets-on-table* & *try-to-load-gun* & *have-bullets* & *have-gun*)

· [(**PROB**(*can-try-to-aim-gun* & *believe-light-is-green* / *try-to-pickup-gun* & *gun-on-table* & *try-to-pickup-bullets* & *bullets-on-table* & *try-to-load-gun* & *have-bullets* & *have-gun* & *try-to-look-at-light* & *can-see-light* & *light-green*)

· {[**PROB**(*gun-aimed* & *gun-loaded* & *can-try-to-fire-gun* / *try-to-pickup-gun* & *gun-on-table* & *try-to-pickup-bullets*

& *bullets-on-table* & *try-to-load-gun* & *have-bullets*
& *have-gun* & *try-to-look-at-light* & *can-see-light*
& *light-green* & *believe-light-is-green* & *try-to-aim-gun*)
· **PROB**(*target-hit* / *try-to-pickup-gun* & *gun-on-table*
& *try-to-pickup-bullets* & *bullets-on-table*
& *try-to-load-gun* & *have-bullets* & *have-gun*
& *try-to-look-at-light* & *can-see-light*
& *light-green* & *believe-light-is-green*
& *try-to-aim-gun* & *try-to-fire-gun* & *gun-aimed*
& *gun-loaded*)]

+ [**PROB**(*gun-aimed* & ~*gun-loaded* & *can-try-to-fire-gun* /
*try-to-pickup-gun* & *gun-on-table* & *try-to-pickup-bullets*
& *bullets-on-table* & *try-to-load-gun* & *have-bullets*
& *have-gun* & *try-to-look-at-light* & *can-see-light*
& *light-green* & *believe-light-is-green* & *try-to-aim-gun*)
· **PROB**(*target-hit* / *try-to-pickup-gun* & *gun-on-table*
& *try-to-pickup-bullets* & *bullets-on-table*
& *try-to-load-gun* & *have-bullets* & *have-gun*
& *try-to-look-at-light* & *can-see-light*
& *light-green* & *believe-light-is-green*
& *try-to-aim-gun* & *try-to-fire-gun* & *gun-aimed*
& ~*gun-loaded*)]

...

This sum contains 6,436 probability terms.

The sum can be greatly simplified if we add some auxiliary probability information. For instance, let us assume:

**PROB**(*can-try-to-aim-gun* / *have-gun*) = 0.8
**PROB**(*can-try-to-load-gun* / (*have-gun* & *have-bullets*)) = 0.9
**PROB**(*can-try-to-load-gun* / ~*have-gun*) = 0
**PROB**(*can-try-to-load-gun* / ~*have-bullets*) = 0
**PROB**(*can-try-to-aim-gun* / ~*have-gun*) = 0
**PROB**(*can-try-to-fire-gun* / *have-gun*) = 1
**PROB**(*can-try-to-fire-gun* / ~*have-gun*) = 0
**PROB**(*have-gun* / ~*gun-on-table*) = 0
**PROB**(*have-bullets* / ~*bullets-on-table*) = 0
**PROB**(*gun-aimed* / ~*can-try-to-aim-gun*) = 0
**PROB**(*gun-loaded* / ~*can-try-to-load-gun*) = 0
**PROB**(*target-hit* / ~*gun-aimed*) = 0
**PROB**(*target-hit* / ~*gun-loaded*) = 0
**PROB**(*can-try-to-pickup-gun* / *gun-on-table*) = 0.9
**PROB**(*can-try-to-pickup-bullets* / *bullets-on-table*) = 0.9
**PROB**(*can-try-to-pickup-gun* / ~*gun-on-table*) = 0
**PROB**(*can-try-to-pickup-bullets* / ~*bullets-on-table*) = 0
**PROB**(*believe-light-is-green* / ~*can-see-light*) = 0

The probabilities with zero values greatly simplify the sum, reducing it to a sum of just 52 probability terms.

If we focus on the remaining probabilities, many of them can be simplified by direct inference. For example, it is reasonable to assume defeasibly that

**PROB**(*target-hit / try-to-pickup-gun & gun-on-table & try-to-pickup-bullets & bullets-on-table & try-to-load-gun & have-bullets & have-gun & try-to-look-at-light & can-see-light & light-green & believe-light-is-green & try-to-aim-gun & try-to-fire-gun & gun-aimed & gun-loaded*)
= **PROB**(*target-hit / try-to-fire-gun & gun-aimed & gun-loaded*).

Finally, it is often reasonable to assume that the unconditional probability of various outcomes is approximately zero. For instance, thinking of all the objects in the world and how many of them are hit by bullets, it is reasonable to estimate that **PROB**(*target-hit*) = 0. If we add that to our probabilistic information, direct inference gives us a simple, easily manageable, sum:

**PROB**(*can-try-to-pickup-gun / gun-on-table*)
· **PROB**(*gun-on-table*)
· **PROB**(*can-try-to-pickup-bullets / bullets-on-table*)
· **PROB**(*bullets-on-table*)
· **PROB**(*can-try-to-load-gun / have-gun & have-bullets*)
· **PROB**(*have-bullets / try-to-pickup-bullets & bullets-on-table*)
· **PROB**(*have-gun / try-to-pickup-gun & gun-on-table*)
· {[**PROB**(*can-see-light / light-green*)
· **PROB**(*light-green*)
· **PROB**(*can-try-to-aim-gun / have-gun*)
· **PROB**(*believe-light-is-green / try-to-look-at-light & can-see-light & light-green*)
· **PROB**(*gun-aimed / try-to-aim-gun*)
· **PROB**(*gun-loaded / try-to-load-gun*)
· **PROB**(*target-hit / try-to-fire-gun & gun-aimed & gun-loaded*)]
+ [**PROB**(*can-see-light / ~light-green*)
· **PROB**(*~light-green*)
· **PROB**(*can-try-to-aim-gun / have-gun*)
· **PROB**(*believe-light-is-green / try-to-look-at-light & can-see-light & ~light-green*)
· **PROB**(*gun-aimed / try-to-aim-gun*)
· **PROB**(*gun-loaded / try-to-load-gun*)
· **PROB**(*target-hit / try-to-fire-gun & gun-aimed & gun-loaded*)]}

The upshot is that what start as outrageously complicated probabilities can be reduced by direct inference to simple probabilities that are easily computed by a resource bounded agent.

## 6. Conclusions

Generalizing our earlier conclusions about action omnipotence, I have argued that plans can be translated into conditional linear policies, with the result that the expected utility of the plan can be identified with the marginal expected value of the corresponding policy. This gives us a precise definition of the expected utility of a plan that can be used in decision-theoretic reasoning. However, expanding the definition of the causal probability of an outcome given such a policy can lead to an immense sum. But it was shown that direct inference can simplify the sum so that the computations become manageable.

# 12
# Locally Global Planning

## 1. The Theory

To repeat the conclusion of chapter 10, I propose that practical cognition should not aim at finding optimal solutions to practical problems. A rational cognizer should instead look for good solutions, and replace them with better solutions if any are found. Solutions come in the form of plans. However, the goodness of a plan cannot be viewed in isolation from all the other plans the agent has adopted. The latter make up the agent's master plan, and practical cognition aims at improving the master plan. In general, a change to the master plan may consist of deleting several local plans and adding several others. Where $M$ is a master plan and $C$ a change, $M\Delta C$ is the result of making the change to $M$. The marginal expected utility of a change $C$ is the difference it makes to the expected utility of the master plan:

$$\mathbf{MEU}(C,M) = \mathbf{EU}(M\Delta C) - \mathbf{EU}(M).$$

The *principle of locally global planning* was then formulated as follows:

> It is rational for an agent to make a change $C$ to the master plan $M$ iff the marginal expected utility of $C$ is positive, that is, iff $\mathbf{EU}(M\Delta C) > \mathbf{EU}(M)$.

This theory is still pretty schematic. It leaves most details to the imagination of the reader, and in philosophy, that is often a recipe for disaster. When philosophical theories fail, it is not because they did not look good in the abstract—it is because it is impossible to make the details work. My ultimate objective is to produce a theory of sufficient precision that it is possible to implement it in an artificial cognitive agent, but this book will stop short of that ideal. So this chapter will simply give a preliminary sketch of how a resource bounded agent can actually perform locally global planning.

## 2. Incremental Decision-Theoretic Planning

This final chapter will make some first, tentative, steps toward implementation. At this point, they are somewhat speculative. The principle of locally global planning forms the basis for an "evolutionary" theory of rational decision making. It is a very general principle. An agent that directs its activities in accordance with this principle will implement the algorithm scheme diagrammed in figure 12.1. However, a complete theory of rational decision making must include more details. In particular, it must include an account of how candidate changes *C* are selected for consideration. This is an essential part of any theory of how rational agents should go about deciding what actions to perform. The selection of candidate changes is a complex matter and all I can do here is sketch a preliminary account, but it is important to at least do that in order to make it credible that locally global planning can provide the basis for rational decision making.

The logically simplest algorithm having the form of figure 12.1 would select potential changes randomly, evaluate them by computing their marginal expected utilities, and on that basis decide whether to make them. This would be a version of the "British Museum algorithm". But like any use of the British Museum algorithm, it is computationally infeasible. There are infinitely many possible changes. A feasible algorithm must select potential changes in some more intelligent fashion, concentrating on changes that have some reasonable chance of being desirable changes. Algorithms that direct the search for possible changes to the master plan are *planning algorithms*. The details of constructing such an algorithm are going to be complex, and there is no hope of working them out completely in the space of this chapter. However, some general remarks may be in order.

Let it be acknowledged immediately that there may be more than one good way of expanding the algorithm scheme of figure 12.1. In deterministic planning, AI planning theory has produced a number of quite different planning algorithms, and the same may be true in decision-theoretic planning. The planning context and the cognitive resource bounds of the agent may determine which algorithms are best, with the result that there may be no single algorithm that is best for all agents and all times. So the remarks I will make here should be viewed as suggestions for one way in which decision-theoretic planning can work. This may not be the only way.

If we reflect upon human planning, it seems to have the general structure of what is called "refinement planning". In refinement planning, the planning agent begins by constructing a crude local plan, then searches for "flaws" in the plan, and refines the plan to remove the flaws. For classical planning, this can be illustrated with an example due to Pednault (1987). Suppose my briefcase is at home, and my paycheck is in my briefcase. I adodpt the conjunctive goal of having my briefcase at the office and my paycheck at home. A classical planner might begin by observing that the paycheck is already at home, so nothing needs to be done to achieve that part of the goal. To get the briefcase to the office, it adopts the plan to take it there. So it produces this simple one-step plan. Then it looks for flaws in the plan, and quickly finds one. Because the paycheck is in the briefcase, taking the briefcase to the office will result in the paycheck no longer being at home. This is called "threat detection". The next step is to try to fix the flaw, and this can be done by removing the paycheck from the briefcase before taking

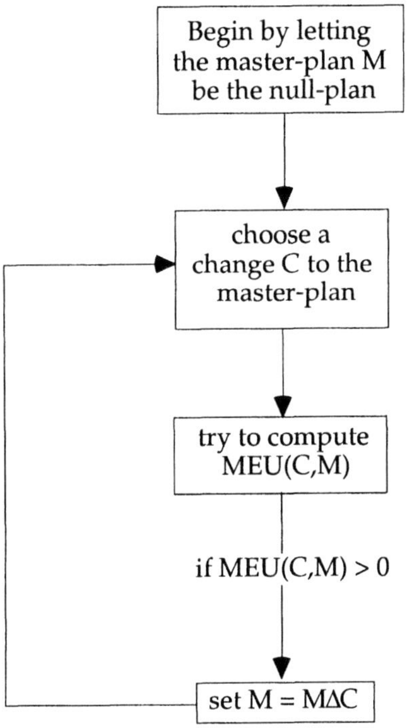


Figure 12.1. An algorithm scheme for locally global planning

the briefcase to the office. This is called "confrontation". In confrontation, we add steps to the plan to remove the threat. The end result is the two-step plan to first remove the paycheck from the briefcase, and then take the briefcase to the office.

In contemporary AI planning theory, refinement planning just refines the plan being constructed, but in locally global planning, refinement planning must be generalized, allowing both the local plan and the master plan to be refined in light of the discovery of destructive interference between the two plans. This can produce multiplan changes to the master plan. It is crucial to the computational feasibility of this general approach that the construction of multiplan changes is driven by the construction of individual local plans. These changes involve newly constructed local plans and sets of previously adopted plans found to interfere with or be interfered with by the new local plan.

My proposal for how the master plan can be constructed and incrementally improved will be based on the account of refinement planning that I will sketch below. But before we do that, it is best to step back and ask why we should expect this general approach to planning to lead to the incremental improvement of the master plan. This can be justified if we make four defeasible assumptions. I will refer to these as the *pivotal planning assumptions*:

**Assumption 1:** The process of constructing "crude local plans" produces plans that will normally have positive expected utilities.

**Assumption 2:** Ordinarily, the expected utility of the result of merging two plans will be the sum of the expected utilities of the two plans.

**Assumption 3:** Computationally feasible reasoning procedures will reveal those cases in which the second assumption fails.

**Assumption 4:** There will be "repair techniques" that can often be used to modify either the local plans or the master plan in such a way as to remove the destructive interference leading to the failure of the second assumption without having to replan from scratch.

The justification of these assumptions will unfold below. Given the pivotal planning assumptions, the planning agent can begin the construction of the master plan by constructing a single local plan having a positive expected utility, and take that to be the master plan. Then the agent can systematically construct further local plans with positive expected utilities, and on the basis of the second assumption it can be assumed defeasibly that each time one of them is merged with the existing master plan, the result will be a master plan with a higher expected utility. On the basis of the third assumption, rational investigation will enable the agent to discover those cases in which the defeasible assumptions fail. This amounts to discovering destructive interference. The fourth assumption tells us that it will often be possible to refine the local plan and/or the master plan so as to avoid the destructive interference, thus leading to a modification of the original plans which, when merged, produces a master plan with a higher expected utility than the original master plan. By proceeding in this way, a rational agent can systematically evolve progressively better master plans.

Now let us turn to the justification of the pivotal planning assumptions. This will simultaneously generate a better understanding of the way in which refinement planning might work in order to make the four assumptions true.

## 3. Goal-Directed Planning

The evolution of the master plan begins with the construction of what I have called "crude local plans". The first pivotal planning assumption is that rational cognition can produce plans that have some reasonable chance of having positive expected utilities and hence of constituting useful additions to the master plan. We have little chance of finding such plans by constructing plans randomly and then assessing their expected utilities. One way to achieve more intelligent plan construction is goal-directed planning. In goal-directed planning, the agent adopts goals, and then searches for ways of achieving them by using goal-regression. In goal-regression, we observe that a goal could probably be achieved in a certain way if a certain condition were satisfied. If the condition is not currently satisfied, we adopt the condition as a new subgoal. In this way, we work backwards from goals to subgoals until we arrive at subgoals that are already satisfied. The details of goal-regression planning are complex, but see my (1998) for a general theory of deterministic goal-regression planning.

Goal-regression is not the only game in town. See Weld (1999) for a discussion of some alternative approaches to goal-directed planning in AI. I am skeptical, however, of being able to apply the current "high-performance" planning algorithms in domains of real-world complexity. This is because those algorithms have stringent knowledge requirements. They make essential use of what is known as "the closed world assumption". This is the assumption that the planning agent knows everything there is to know about the world (at least expressible in the vocabulary of the planning problem). That is totally unreasonable outside of toy problems. For example, if I want to adopt a kitten, in order to plan how to do that using any of these planners I would have to have explicit knowledge of every kitten in the world and all of its properties that are relevant to adoptability. Such an assumption is simply a non-starter for real-world planning. But it is absolutely essential to planners like GRAPHPLAN (Blum and Furst 1995, 1997) and BLACKBOX (Kautz and Selman 1996, 1998). There is no way to make them work without it.

Goals are simply valuable states of affairs—states of affairs that, were they achieved, would add to the overall value of the world (from the agent's point of view). Any valuable state of affairs can be chosen as a goal. In the cognitive architecture of a rational agent, that has the effect of initiating the search for plans for achieving the goal. The agent will probably not be in a position to construct plans for achieving most valuable states of affairs, so although they may technically be regarded as goals, they will have no effect on the agent's reasoning. They will simply be recorded as desirable, and then left alone unless the agent later encounters a way of achieving them.

The significance of goal-directed planning is that it produces local plans for the achievement of valuable goals. Assuming that the planning process will monitor the cumulative execution costs of the plan steps and not produce plans with high execution costs for lower-valued goals, goal-directed planning will produce plans that it is defeasibly reasonable to expect to have positive expected utilities. Then in accordance with the second pivotal planning assumption, it is defeasibly reasonable for the agent to expect to be able to simply add the new plans to the master plan and thereby improve the master plan.

Much work in AI planning theory has focused on goal-directed planning. However, this is not adequate as a general account of planning. This is most easily seen by thinking more carefully about goals. As I have described them, goals are "goals of attainment". That is, they are valuable states of affairs that can, in principle be brought about, thereby attaining the goal. Rational agents operating in an environment of real world complexity may continually acquire new goals of attainment in response to the acquisition of new knowledge about their environment. If I learn that my favorite author has written a new book, I may acquire the goal of reading it. If I learn that a tiger is stalking me, I may acquire the goal of escaping from the tiger. If I learn that my fuel level is running low, I may acquire the goal of getting more fuel. Note that these are goals that I cannot have until I acquire the requisite knowledge.

It is important to realize that these new goals of attainment are not simply instances of more general goals of attainment that I had all along. For example, my desire not to run out of fuel when I realize that I am in danger of doing so derives from a general desire not to run out of fuel. But that general desire cannot be represented as a goal of attainment. In particular, it is not the goal of *never running out of fuel*. If that were my goal, then if I ever did run out of fuel that goal would become henceforth unachievable regardless of anything I might do in the future, and so there would be no reason in the future for trying to avoid running out of fuel. My general desire is more like a *disposition* to form goals. As such, it is not itself the target of planning.[1] I suggest that such general desires are the same thing as feature likings, in the sense of chapter 3.

In different circumstances, an agent may be presented with different opportunities or exposed to different dangers. An opportunity is an opportunity to achieve an outcome having positive utility. This is accomplished by adopting the achievement of that outcome as a goal and then planning accordingly. Similarly, a danger is a danger that some outcome of negative utility will occur. That should inspire the agent to adopt the goal of preventing that outcome, and planning accordingly.

Planning for such newly acquired goals is not different in kind from planning for fixed goals, so this creates no problem for a theory of goal-directed planning. What is harder to handle, however, is the observation that we often plan ahead for "possible dangers" or "possible opportunities", without actually knowing that they will occur. For example, while trekking through tiger country, I may note that I might encounter a tiger, observe that if I do I will acquire the goal of not being eaten by him, and with that in mind I will plan to carry a gun. The important point here is that I am planning for a goal that I do not yet have, and I may begin the execution of the plan before acquiring the goal (that is, I will set out on my trek with a gun).

These are examples of what, outside of AI, is usually called "contingency planning".[2] We plan for something that *might* happen, without knowing whether it actually will happen. If it does happen, then there will be a goal of the sort goal-directed planning normally aims at achieving. What makes

---

[1] My general desire might be represented as the desire to run out of fuel as infrequently as possible, but that is not a goal of attainment — there is no point at which I can be said to have achieved that goal.

[2] In AI, the term "contingency planning" is used instead to talk about plans with conditional steps.

it reasonable to begin planning before we actually acquire the information that generates the goals—just on the promise that we might acquire the information—is that such contingency plans can have positive expected utilities even before we acquire the information. So in decision-theoretic planning, such contingency planning is desirable, but the problem is to fit it into the framework of our planning theory. The goals are hypothetical, but the plans are real in the sense that we may not only adopt them but also start executing them before acquiring the goals at which they aim. Boutelier et al. (1999) make a similar observation, taking it to be a criticism of goal-directed planning.

The best way to understand contingency planning is to think more generally about the function that goals play in planning. The rational agent is trying to find plans with positive expected utilities. Goals are of only instrumental value in this pursuit. On pain of computational infeasibility, we cannot use the British Museum algorithm and randomly survey plans until we find plans with high expected utilities. We must direct our planning efforts in some more intelligent fashion, and goals provide a mechanism for doing that. Goals are features of high value, so there is a defeasible presumption that a plan that achieves a goal will have positive expected utility. Hence goal-directed planning is a mechanism for finding plans with positive expected utilities. But what these examples show is that goal-directed planning by itself will not always suffice for finding such plans. It is too restrictive to require that planning must always be initiated by actual goals. It must be possible to plan for hypothetical goals as well. Hypothetical goals will become goals *if* something is the case. Escaping from the tiger will become a goal if I learn there is a tiger, and acquiring more fuel will become a goal if I learn that I am running low on fuel. To initiate planning for these hypothetical goals, it should suffice to know that the antecedent is sufficiently probable to make a plan for achieving the goal have a positive expected utility. What we can do is take the value of the goal in the context of the planning to be the value the goal would have were the antecedent true, discounted by the probability the antecedent will be true. Then we can engage in planning "as if" the goal were a real goal, and evaluate the plan in terms of the attenuated value of the goal. If the resulting plan has a sufficiently high marginal expected utility, we can add it to the master plan and begin its execution before the goal becomes real (for example, we can carry a gun or top off the fuel tank).

The preceding remarks constitute only the briefest sketch of some aspects of goal-directed planning, but hopefully they will point the way to a general theory of goal-directed planning that can be incorporated into a more mature theory of locally global planning. These remarks are intended as a sketchy defense of the first pivotal planning assumption. I do not suppose for a moment that it is going to be an easy task to work out a theory of goal-directed planning for a decision-theoretic planner. All I suggest is that it may be useful to try. One thing is worth noting. Classical (non-decision-theoretic) goal-directed planners have difficulty solving hard problems that can often be solved by the kind of high-performance planners mentioned above. As remarked above, they are, unfortunately, inapplicable to the planning of autonomous agents in the real world, because of their reliance on the closed world assumption, so for general planning we seem to be stuck with some form of goal-directed planning. If planning problems are modeled as systematic and blind search through a tree of actions, the problem quickly becomes computationally too difficult. However, the problem may actually be easier in decision-theoretic contexts, because the search need not be blind. We have important information at our disposal for guiding search that we do not have in classical planning, namely probabilities and utilities. How much this helps remains to be seen, but I think it is at least plausible to suppose it will help a lot.

## 4. Presumptively Additive Expected Utilities

The second pivotal planning assumption was that, ordinarily, the expected utility of the result of merging two plans will be the sum of the expected utilities of the two plans. It follows from this that we can assume defeasibly that the marginal expected utility of a plan is equal to its expected utility. In other words, when we add a new local plan to the master plan, the result is to increase the expected utility of the master plan by an amount equal to the expected utility of the local plan. It is this assumption that makes it feasible to try to improve the master plan incrementally, by constructing local plans and adding them to it.

Let *M* be the master plan, and *P* be a local plan we are considering merging with *M* to produce the merged plan *M*+*P*. Both *M* and *M*+*P* will normally be nonlinear plans. The expected utility of a nonlinear plan is determined by the expected utilities of its linearizations. If a planning agent discovers that some linearizations have lower expected utilities than others, then he should add ordering constraints that preclude those linearizations. So, as in chapter 11, it is reasonable to define the expected utility of a nonlinear plan to be the maximum of the expected utilities of its linearizations. Ideally, all the linearizations should have the same expected utility. A linearization *L* of *M*+*P* will result from inserting the steps of a linearization of *P* into a linearization of *M*. The expected utility of *L* will be the expected utility of *M* plus the expected utility of *P in the context of L*. By the latter, I mean that the probabilities and utilities employed in computing the expected utility of *M* or *P* in that context will all be conditional on the previous steps of *L* having been attempted.

If it is defeasibly reasonable to expect the utilities and probabilities of elements of *P* to remain unchanged when they are made conditional on the previous steps of *L* having been attempted, it will follow that it is defeasibly reasonable to expect that **EU**(*L*) = **EU**(*M*) + **EU**(*P*), and hence **EU**(*M*+*P*) = **EU**(*M*) + **EU**(*P*). This divides into two separate expectations—that the utilities remain unchanged, and that the probabilities remain unchanged. In earlier chapters, I have argued that both of these expectations are defeasibly reasonable. I will briefly rehearse the arguments for these claims.

Recall from chapter 7 that direct inference is the kind of inference involved in deriving definite (single-case) probabilities from indefinite (general) probabilities. For instance, it governs the kind of inference involved in inferring the probability that it will rain today from general probabilities of rain in different meteorological conditions. I proposed a general theory of direct inference in my (1990), and it is sketched briefly in chapter 7 and more fully in the appendix. Causal probabilities were defined in chapter 8, and I showed that the general principles underlying direct inference imply a defeasible presumption of statistical irrelevance for causal probabilities:

(CIR) For any $P,Q,R$, it is defeasibly reasonable to expect that
$$\textbf{C-PROB}_A(P/Q\&R) = \textbf{C-PROB}_A(P/Q).$$

This is exactly the principle we need to justify the defeasible assumption that the probabilities relevant to the computation of the expected utility of a plan do not change when the plan is merged with the master plan. Of course, the identity in (CIR) will often fail in concrete cases, but the point of (CIR) is that it is reasonable to expect it to hold unless we have a definite reason for thinking otherwise.

Turning to utilities, there can be no logical guarantee that conditional utilities do not vary with context. If this happened too often, we would be unable to compute utility-measures for complex combinations of parameters on the basis of the utility-measures of the individual parameters or small combinations of them, and that in turn would make decision-theoretic reasoning impossible. There must be at least a defeasible presumption that for any state of affairs $P$ and circumstances $C$, $\mathbf{U}(P) = \mathbf{U}(P/C)$. Such a defeasible presumption follows from the principles of causal reasoning used in chapter 5 to defend the database calculation. It was argued that it is defeasibly reasonable to expect disparate events to be causally independent unless we have some concrete reason for thinking otherwise. This was made precise in terms of the "principle of causal irrelevance" and the "principle of causal independence". The concept of value employed in this book is a causal one. Appealing to chapter 3, the abstract value of a feature is a measure of its tendency to cause state liking. This has the consequence that general principles for reasoning about values can be derived from analogous principles for reasoning about causation. The principle of causal irrelevance has the consequence that it is defeasibly reasonable to expect features to be value-neutral, that is, to cause no change in the quantity of state liking produced by another feature. In other words, $\mathbf{U}(P/C) = \mathbf{U}(P)$. This is precisely the assumption that we need regarding the utilities that play a role in computing $\mathbf{EU}(M+P)$.

The second pivotal planning assumption follows from these two principles for reasoning defeasibly about probabilities and utilities. So I regard it as established.

## 5. Finding and Repairing Decision-Theoretic Interference

The third pivotal planning assumption is that computationally feasible reasoning will reveal those cases in which the second assumption fails. The second assumption was that when two plans are merged into a single plan, the expected utility of that composite plan will be the sum of the expected utilities of the constituent plans. When this assumption holds, let us say that the plans exhibit *decision-theoretic independence*. *Decision-theoretic interference* is the failure of decision-theoretic independence. In adding local plans to the master plan, our defeasible assumption is one of decision-theoretic independence, so what is needed is tools for detecting decision-theoretic interference.

Consider the ways in which decision-theoretic interference can arise. The expected utility of a plan is determined by various probabilities and utilities. The probabilities are (1) the probabilities that goals will be achieved, side effects will occur, or execution costs will be incurred if certain combinations of plan steps are attempted, and (2) the probabilities that the agent can attempt to perform a plan step if certain combinations of earlier plan steps are attempted and certain subgoals are achieved. The utilities are the values of goals, side-effects, and execution costs conditional on certain combinations of plan steps having been attempted and certain subgoals having been achieved. Decision-theoretic interference arises from embedding the plan in a larger context (the master plan) in which some of these probabilities or utilities change. Without going into detail, it is clear that algorithms can be designed for searching for decision-theoretic interference by looking for constituents of the master plan that will change these probabilities and utilities. For example, if a plan relies upon the probability **PROB**(*goal*/*subgoal* & action $A$ is performed), we can search for "underminers" $P$ such that **C-PROB**$_A$(*goal*/*subgoal* & $P$) ≠ **C-PROB**$_A$(*goal*/*subgoal*) and such that the master plan contains a subplan that achieves $P$ with some probability. This will be analogous to threat detection in deterministic goal-regression planning.

The fourth and final pivotal planning assumption is that we will often be able to make relatively small changes to local plans or the master plan to avoid any decision-theoretic interference that is detected. This is analogous to what occurs in classical refinement planning, and many of the same repair techniques will be applicable. The simplest will consist of adding ordering constraints to $M$+$P$ to avoid interference. Another way of repairing the decision-theoretic interference is to add further steps that prevent the lowering of the probability. This is analogous to what is called *confrontation* in classical planning (Penberthy and Weld 1992). The upshot is that familiar ideas taken from classical planning will be applicable here as well. There may also be other ways of resolving decision-theoretic interference that do not correspond to techniques used in classical planning. That is a matter for further research. But this much is clear. The discovery of decision-theoretic interference need not cause us to reject our local plans altogether. We may instead be able to modify them in small ways to resolve the interference. Of course, this will not *always* be possible. Sometimes the interference will be irresolvable, and then we must reject one or more local plans and look for other ways of achieving some of our goals. However, there is no reason to expect this to make decision-theoretic planning intractable for realistic (resource bounded) rational agents.

## 6. Conclusions

We have reached the end of a long path. Our topic has been rational decision making in real agents (human or artificial), with all of their cognitive limitations. I began with the optimality prescription, according to which decision makers are supposed to select actions whose expected values are at least as great as that of any competitors. To make the optimality prescription precise, Part I undertook an investigation of the values used in computing expected values, and Part II undertook an investigation of the probabilities.

It was argued in Part I that the utilities employed in decision-theoretic reasoning should be regarded as *abstract values*. The abstract value of a feature (situation type) was defined to be its tendency to contribute to the agent's state liking over time, that is, the mathematical expectation of the cumulative state liking caused by the feature. Numerous problems arise in trying to apply this to human beings. They stem from the fact that abstract values will often be hard to compute. It was suggested that these problems are solved in human beings by equipping them with a number of Q&I modules enabling them to compute utilities and expected values approximately.

It was argued in Part II that subjective probability makes no sense when our subject is real agents rather than ideal agents. Decision making must be based on beliefs about non-subjective probabilities. It was argued that the requisite kind of probability for use in decision making is a causal probability defined ultimately in terms of nomic probability.

Parts I and II equipped us with a precise version of the optimality principle. Then we began its evaluation. It was argued in chapter 9 that expected value is not the correct measure to use for evaluating actions. The difficulty is that it ignores the fact that we may be uncertain to varying degrees whether we can perform an action. This difficulty was resolved by replacing the appeal to expected values by an appeal to expected utilities, where the latter were defined to be the marginal expected values of *conditional policies* of the form *try-A if can-try-A*. This produces a revised version of the optimality prescription that still retains the spirit of the original.

In chapter 10, however, it was argued that the optimality prescription is fundamentally flawed. The difficulty is twofold. First, actions cannot be evaluated in isolation. They must be evaluated in the context of all the other actions the agent plans to perform. This led me to the conclusion that the proper object of decision-theoretic evaluation is *plans* rather than actions. The oft-heard response to this conclusion is that we should simply apply the optimality prescription to plans rather than actions. But now the second difficulty arises. First, if we confine our attention to local plans, there is no way to define "alternative plans" so that rational choice consists of choosing an optimal plan from a list of alternatives. Local plans cannot be compared meaningfully in terms of their expected utilities. Second, if we look instead at universal plans, real decision makers would not be able to construct optimal universal plans (even if they exist), both because universal plans are too complex for real agents to construct and because finding optimal ones would require surveying and comparing infinitely many plans.

The fundamental idea behind the optimality prescription is that decision-theoretic reasoning should be aimed at finding optimal solutions to practical problems, and that is what must be rejected. There may be no optimal plans, and even if there were, real decision makers would not be able to find them. What rationality must instead prescribe is the search for good plans, with the understanding that if better plans are found later they will be used to replace the previously adopted plans.

But how is "good plan" to be understood if plans cannot be evaluated in terms of their expected utilities? My proposal is that plans must be evaluated in the context of all the other plans the agent has adopted. We can merge all of these plans into a single plan, called *the master plan*. Then we can view the objective of practical cognition as that of improving the master plan. New plans are adopted and old plans withdrawn because by doing so we can increase the expected utility of the master plan. Planning becomes a non-terminating process of systematically improving the master plan rather than a terminating search for optimal solutions.

Master plans are potentially very large, combining everything an agent plans to do. Can we regard realistic resource-bounded agents as evaluating and attempting to improve their master plans? The theory of *locally global planning*, sketched in this chapter, is an attempt to show how real agents can actually do this. The key is to employ various principles (discussed in chapter 11) that enable them to reason defeasibly about plans without having to work out the explicit (and immense) sums that result from expanding the definition of "expected utility" when it is applied to master plans. This chapter has been no more than a sketch. Much work remains to be done, and I will not be confident that I have the details right until I can build an implemented AI system that performs such reasoning automatically. That is the aim of ongoing research in the OSCAR project.[3]

[3] Up-to-date information about the OSCAR project is available online at http://oscarhome.soc-sci.arizona.edu/ftp/OSCAR-web-page/oscar.html.

# Appendix: The Theory of Nomic Probability

## 1. Introduction

The purpose of this appendix is to give a sketch of some of the more technical parts of the theory of nomic probability. Chapter 7 gave a very brief sketch. This appendix expands that sketch. But this is still only a sketch. For the complete theory, the reader should consult my book *Nomic Probability and the Foundations of Induction* (1990).

There are two kinds of physical laws—statistical and nonstatistical. Statistical laws are probabilistic. The kind of probability involved in statistical laws is what I call *nomic probability.* The best way to understand nomic probability is by looking first at non-statistical laws. What distinguishes such laws from material generalizations of the form "$(\forall x)(Fx \rightarrow Gx)$" is that they are not just about actual $F$'s. They are about "all the $F$'s there could be"—that is, they are about "physically possible $F$'s". I call non-statistical laws *nomic generalizations.* Nomic generalizations can be expressed in English using the locution "Any $F$ would be a $G$". I will symbolize this nomic generalization as "$Fx \Rightarrow Gx$". It can be roughly paraphrased as telling us that any physically possible $F$ would be $G$. *Physical possibility*, symbolized "$\Diamond_p Q$", means that $Q$ is logically consistent with the set of all true nomic generalizations.

I propose that we think of nomic probabilities as analogous to nomic generalizations. Just as "$Fx \Rightarrow Gx$" tells that us any physically possible $F$ would be $G$, for heuristic purposes we can think of the statistical law "$\mathbf{prob}(Fx/Fx) = r$" as telling us that the proportion of physically possible $F$'s that would be $G$'s is $r$. For instance, pretend it is a law of nature that at any given time, there are exactly as many electrons as protons. Then in every physically possible world, the proportion of electrons-or-protons that are electrons is 1/2. It is then reasonable to regard the probability of a particular particle being an electron given that it is either an electron or a proton as 1/2. Of course, in the general case, the proportion of $F$'s that are $G$'s will vary from one possible world to another. $\mathbf{prob}(Gx/Fx)$ then "averages" these proportions across all physically possible worlds. The mathematics of this averaging process is complex, and I will say more about it below.

Nomic probability is illustrated by any of a number of examples that are difficult for frequency theories. For instance, consider a physical description $D$ of a coin, and suppose there is just one coin of that description and it is never flipped. On the basis of the description $D$ together with our knowledge of physics we might conclude that a coin of this description is a fair coin, and hence the probability of a flip of a coin of description *D* landing heads is 1/2. In saying this we are not talking about relative frequencies—as there are no flips of coins of description *D*, the relative frequency does not exist. Or suppose instead that the single coin of description *D* is flipped just once, landing heads, and then destroyed. In that case the relative frequency is 1, but we would still insist that the probability of a coin of that description landing heads is 1/2. The reason for the difference between the relative frequency and the probability is that the probability statement is in some sense subjunctive or counterfactual. It is not just about actual flips, but about possible flips as well. In saying that the probability is 1/2, we are saying that out of all physically possible flips of coins of description *D*, 1/2 of them would land heads. To illustrate nomic probability with a more realistic example, in physics we often want to talk about the probability of some event in simplified circumstances that have never occurred. The typical problem given students in a quantum mechanics class is of this character. The relative frequency does not exist, but the nomic probability does and that is what the students are calculating.

The theory of nomic probability will be a theory of probabilistic reasoning. I will not attempt to *define* "nomic probability" in terms of simpler concepts, because I doubt that can be done. If we have learned anything from twentieth century philosophy, it should be that philosophically interesting concepts are rarely definable. You cannot solve the problem of perception by defining "red" in terms of "looks red", you cannot solve the problem of other minds by defining "person" in terms of behavior, and you cannot provide foundations for probabilistic reasoning by defining "probability" in terms of relative frequencies. In general, the principles of reasoning involving a concept are primitive constituents of our conceptual framework and cannot be derived from definitions. The task of the epistemologist must simply be to state the principles precisely. That is my objective for probabilistic reasoning.

Probabilistic reasoning has three components. First, there must be rules prescribing how to ascertain the numerical values of nomic probabilities on the basis of observed relative frequencies. Second, there must be computational principles that enable us to infer the values of some nomic probabilities from others. Finally, there must be principles enabling us to use nomic probabilities to draw conclusions about other matters.

The first element of this account consists largely of a theory of statistical induction. The second element will consist of a calculus of nomic probabilities. The final element is an account of how conclusions not about nomic probabilities can be inferred from premises about nomic probability. This has two parts. First, it seems clear that under some circumstances, knowing that certain probabilities are high can justify us in holding related non-probabilistic beliefs. For example, I know it to be highly probable that the date appearing on a newspaper is the correct date of its publication. (I do not know that this is always the case—typographical errors do occur). On this basis, I can arrive at a justified belief regarding today's date. The epistemic rules describing when high probability can justify belief are called *acceptance rules*. The acceptance rules endorsed by the theory of nomic probability constitute the principal novelty of that theory. The other fundamental principles that are adopted as primitive assumptions about nomic probability are all of a computational nature. They concern the logical and mathematical structure of nomic probability, and amount to nothing more than an elaboration of the standard probability calculus. It is the acceptance rules that give the theory its unique flavor and constitute the main epistemological machinery making it run.

It is important to be able to make another kind of inference from nomic probabilities. We can make a distinction between "definite" probabilities and "indefinite" probabilities. A definite probability is the probability that a particular proposition is true. Indefinite probabilities, on the other hand, concern properties rather than propositions. For example, the probability of a smoker getting cancer is not about any particular smoker. Rather, it relates the property of being a smoker and the property of getting cancer. Nomic probabilities are indefinite probabilities. This is automatically the case for any probabilities derived by induction from relative frequencies, because relative frequencies relate properties. But for many practical purposes, the probabilities we are really interested in are definite probabilities. We want to know how probable it is that it will rain today, that Bluenose will win the third race, that Sally will have a heart attack, and so on. It is probabilities of this sort that are involved in practical reasoning. Thus the first three elements of our analysis must be augmented by a fourth element. That is a theory telling us how to get from indefinite probabilities to definite probabilities. We judge that there is a 20% probability of rain today because the indefinite probability of its raining in similar circumstances is believed to be about .2. We think it unlikely that Bluenose will win the third race because he has never finished above seventh in his life. We judge that Sally is more likely than her sister to have a heart attack because Sally smokes like a furnace and drinks like a fish, while her sister is a tea-totaling non-smoker who jogs and lifts weights. We take these facts about Sally and her sister to be relevant because we know that they affect the indefinite probability of a person having a heart attack. That is, the indefinite probability of a person who smokes and drinks having a heart attack is much greater than the indefinite probability for a person who does not smoke or drink and is in good physical condition. Inferences from indefinite probabilities to definite probabilities are called *direct inferences.* A satisfactory theory of nomic probability must include an account of direct inference.

To summarize, the theory of nomic probability will consist of (1) a theory of statistical induction, (2) an account of the computational principles allowing some probabilities to be derived from others, (3) an account of acceptance rules, and (4) a theory of direct inference.

## 2. Computational Principles

It might seem that the calculus of nomic probabilities should just be the classical probability calculus. But this overlooks the fact that nomic probabilities are indefinite probabilities. Indefinite probabilities operate on properties and relations. This introduces logical relationships into the theory of nomic probability that are ignored in the classical probability calculus. One simple example is the "principle of individuals":

(IND) $\mathbf{prob}(Axy \;/\; Rxy \;\&\; y{=}b) = \mathbf{prob}(Axb/Rxb)$.

This is an essentially relational principle and is not even expressible in the classical probability calculus. It might be wondered how there can be general truths regarding nomic probability that are not theorems of the classical probability calculus. The explanation is that, historically, the probability calculus was devised with definite probabilities in mind. The standard versions of the probability calculus originate with Kolmogorov (1933) and are concerned with "events". The relationship between the calculus of indefinite probabilities and the calculus of definite probabilities is a bit like the relationship between the predicate calculus and the propositional calculus. Specifically, there are principles regarding relations and quantifiers that must be added to the classical probability calculus to obtain a reasonable calculus of nomic probabilities.

In developing the calculus of nomic probabilities, I propose that we make further use of the idea that nomic probability measures proportions among physically possible objects. The statistical law "$\mathbf{prob}(Gx/Fx) = r$" can be regarded as telling us that the proportion of physically possible *F*'s that would be *G* is *r*. Treating probabilities in terms of proportions proves to be a useful approach for investigating the logical and mathematical structure of nomic probability.

Proportions operate on sets. Given any two sets *A* and *B* we can talk about the proportion of members of *B* that are also members of *A*. I will symbolize "the proportion of members of *B* that are in *A*" as "$\rho(A/B)$". $\rho(A/B)$ is the *relative measure* of *A* in *B*. The concept of a proportion is a general measure-theoretic notion. The theory of proportions was developed in detail in my (1987) and (1990), and I will say more about it below. But first consider how we can use it to derive computational principles governing nomic probability. The derivation is accomplished by making more precise our explanation of nomic probability as measuring proportions among physically possible objects. Where *F* and *G* are properties and *G* is not counterlegal (that is, it is physically possible for there to be *G*'s), we can regard $\mathbf{prob}(Fx/Gx)$ as the proportion of physically possible *G*'s that would be *F*'s. This forces us to consider more carefully just what we mean by "a physically possible *F*". We cannot just mean "a possible object that is *F* in some physically possible world", because the same object can be *F* in one physically possible world and non-*F* in another. Instead, I propose to understand a physically possible *F* to be an ordered pair $\langle w,x\rangle$ such that *w* is a physically possible world (that is, one having the same physical laws as the actual world) and *x* is an *F* at *w*. We then define:

(2.1) $\mathbb{F} = \{\langle w,x\rangle \mid w$ is a physically possible world and $x$ is $F$ at $w\}$;

$$\mathbf{G} = \{\langle w,x\rangle \mid w \text{ is a physically possible world and } x \text{ is } G \text{ at } w\}.$$

With this understanding, we can regard nomic probabilities as measuring proportions between sets of physically possible objects. Given a suitable proportion function ρ we could stipulate that:

(2.2) If $\Diamond_p(\exists x)Gx$ then $\mathbf{prob}(Fx/Gx) = \rho(\mathbf{F}/\mathbf{G})$

where $\mathbf{F}$ is the set of all physically possible $F$'s and $\mathbf{G}$ is the set of all physically possible $G$'s. However, I doubt that we can pick out the right proportion function without appealing to prob itself, so the postulate is simply that *there is* some proportion function related to prob as above. This tells us a great deal about the formal properties of prob. Rather than axiomatizing prob directly, we can adopt axioms for the proportion function. (2.2) must be extended to include the case of counterlegal probabilities, but I will not go into that here.

The next task is to investigate the theory of proportions. That investigation was carried out in my (1987) and (1990) and generates a calculus of proportions that in turn generates a calculus of nomic probabilities. The simplest and least problematic talk of proportions concerns finite sets. In that case proportions are just relative frequencies. Taking $\#X$ to be the cardinality of a set $X$, relative frequencies are defined as follows:

(2.3) If $X$ and $Y$ are finite and $Y$ is nonempty then $\text{freq}[X/Y] = \#(X \cap Y)/\#Y$.

We then have the *Frequency Principle*:

(2.4) If $X$ and $Y$ are finite and $Y$ is nonempty then $\rho(X/Y) = \text{freq}[X/Y]$.

But we also want to talk about proportions among infinite sets. The concept of a proportion in such a case is an extension of the concept of a frequency. The simplest laws governing proportions are those reflected by the classical probability calculus, and they can be axiomatized as follows:

(2.5) $0 \leq \rho(X/Y) \leq 1$.

(2.6) If $Y \subseteq X$ then $\rho(X/Y) = 1$.

(2.7) If $Z \neq \varnothing$ and $Z \cap X \cap Y = \varnothing$ then $\rho(X \cup Y/Z) = \rho(X/Z) + \rho(Y/Z)$.

(2.8) $\rho(X \cap Y/Z) = \rho(X/Z) \cdot \rho(Y/X \cap Z)$.

Given the theory of proportions and the characterization of nomic probabilities in terms of proportions of physically possible objects, we can derive a powerful calculus of nomic probabilities. Much of the theory is rather standard looking. For example, the following versions of the standard axioms for the probability calculus follow from (2.5)–(2.8):

(2.9) $0 \leq \mathbf{prob}(Fx/Gx) \leq 1$.

(2.10) If $(Gx \Rightarrow Fx)$ then $\mathbf{prob}(Fx/Gx) = 1$.

(2.11) If $\Diamond$H and $[Hx \Rightarrow \sim(Fx\&Gx)]$ then
$\mathbf{prob}(Fx \vee Gx/Hx) = \mathbf{prob}(Fx/Hx) + \mathbf{prob}(Gx/Hx)$.

(2.12) If $\Diamond_p(\exists x)$Hx then $\mathbf{prob}(Fx\&Gx/Hx) = \mathbf{prob}(Fx/Hx) \cdot \mathbf{prob}(Gx/Fx\&Hx)$.

The theory of proportions resulting from (2.4)–(2.8) might be termed "the Boolean theory of proportions", because it is only concerned with the Boolean operations on sets. In this respect, it is analogous to the propositional calculus. However, in $\rho(X/Y)$, $X$ and $Y$ might be sets of ordered pairs. That is, they might be relations. There are a number of principles that ought to hold in that case but are not contained in the Boolean theory of proportions. The classical probability calculus takes no notice of relations. For example, the following *Cross-Product Principle* would seem to be true:

(2.13) If $C \neq \varnothing$, $D \neq \varnothing$, $A \subseteq C$, and $B \subseteq D$, then $\rho(A \times B/C \times D) = \rho(A/C) \cdot \rho(B/D)$.

In the special case in which $A$, $B$, $C$ and $D$ are finite, the cross-product principle follows from the frequency principle, but in general it is not a consequence of the classical probability calculus.

To avoid possible misunderstanding, it is important to observe that the cross-product principle for proportions does not entail the following principle regarding *probabilities*:

$$\mathbf{prob}(Ax\&By/Cx\&Dy) = \mathbf{prob}(Ax/Cx) \cdot \mathbf{prob}(Bx/Dx).$$

This principle is clearly false, because the $C$'s and the $D$'s need not be independent of one another. But the analogous principle for proportions is clearly true. For example, when $A$, $B$, $C$, and $D$ are finite this principle is an immediate consequence of the fact that if $A$ has $n$ members and $B$ has $m$ members then $A \times B$ has $nm$ members.

The fact that the cross-product principle is not a consequence of the classical probability calculus demonstrates that the probability calculus must be strengthened by the addition of some "relational" axioms in order to axiomatize the general theory of proportions. The details of the choice of relational axioms turn out to be rather complicated, and I will not pursue them further here. However, the theory developed in my (1990) turns out to have some important consequences. One of these concerns probabilities of probabilities. On many theories, there are difficulties making sense of probabilities of probabilities, but there are no such problems within the theory of nomic probability. The function "**prob**" can relate any two properties, including properties defined in terms of nomic probabilities. In the theory of nomic probability we get:

(PPROB) If $\Diamond[(\exists x)Gx \ \& \ \mathbf{prob}(Fx/Gx) = r]$ then
$\mathbf{prob}\big(Fx \ / \ Gx \ \& \ \mathbf{prob}(Fx/Gx) = r\big) = r.$

One of the most important theorems of the calculus of nomic probabilities is the *Principle of Agreement*, which I will now explain. This theorem follows from an analogous principle regarding proportions, and I will begin by explaining that principle. First note a rather surprising combinatorial fact (at least, surprising to the uninitiated in probability theory). Consider the proportion of members of a finite set $B$ that are in some subset $A$ of $B$. Subsets of $B$ need not exhibit the same proportion of $A$'s, but it is a striking fact of set theory that subsets of $B$ *tend* to exhibit *approximately* the same proportion of $A$'s as $B$, and both the strength of the tendency and the degree of approximation improve as the size of $B$ increases. More precisely, where "$x \approx_\delta y$" means "$x$ is approximately equal to $y$, the difference being at most $\delta$", the following is a theorem of set theory:

(2.14) For every $\varepsilon,\delta > 0$, there is an $n$ such that if $B$ is a finite set containing at least $n$ members then $\mathrm{freq}\big[\mathrm{freq}[A/X] \approx_\delta \mathrm{freq}[A/B] \ / \ X \subseteq B\big] > 1\text{-}\varepsilon$.

It seems inescapable that when $B$ becomes infinite, the proportion of subsets agreeing with $B$ to any given degree of approximation should become 1. This is *The Principle of Agreement for Proportions*:

(2.15) If $B$ is infinite and $\rho(A/B) = p$ then for every $\delta > 0$,
$\rho\big(\rho(A/X) \approx_\delta p \ / \ X \subseteq B\big) = 1.$

This principle seems to me to be undeniable. It is simply a generalization of (2.14) to infinite sets. It is shown in my (1987) and (1990) that this principle can be derived within a sufficiently strong theory of proportions.

The importance of the Principle of Agreement For Proportions is that it implies a Principle of Agreement For Nomic Probabilities. Let us say that $H$ is a *subproperty* of $G$ iff $H$ nomically implies $G$:

(2.16) $H \preceq G$ iff $\Box_p(\forall x)(Hx \rightarrow Gx)$.[1]

The following turns out to be an easy consequence of the Principle of Agreement for Proportions:

(AGREE) If $F$ and $G$ are properties and there are infinitely many physically possible $G$'s and $\mathbf{prob}(Fx/Gx) = p$, then for every $\delta > 0$,
$\mathbf{prob}\big(\mathbf{prob}(F/X) \approx_\delta p \ / \ X \preceq G\big) = 1.$

This is *The Principle of Agreement for Probabilities*. (AGREE) lies at the heart of

---

[1] This can be generalized in the obvious way to relational properties.

the theory of direct inference, and that makes it fundamental to the theory of statistical and enumerative induction.

## 3. The Statistical Syllogism

Rules telling us when it is rational to believe something on the basis of high probability are called *acceptance rules.* The philosophical literature contains numerous proposals for acceptance rules, but most proceed in terms of definite probabilities rather than indefinite probabilities. There is, however, an obvious candidate for an acceptance rule that proceeds in terms of nomic probability. This is the *Statistical Syllogism*, whose traditional formulation is something like the following:

> Most *A*'s are *B*'s.
> This is an *A*.
> ________________
> Therefore, this is a *B*.

It seems clear that we often reason in roughly this way. For instance, on what basis do I believe what I read in the newspaper? Certainly not that everything printed in the newspaper is true. No one believes that. But I do believe that *most* of what is printed in the newspaper is true, and that justifies me in believing individual newspaper reports.

"Most *A*'s are *B*'s" can have different interpretations. It may mean simply that most actual *A*'s are *B*'s. But at least sometimes, "most" statements can be cashed out as statements of nomic probability. On that construal, "Most *A*'s are *B*'s" means "**prob**$(Bx/Ax)$ is high". This suggests the following acceptance rule, which can be regarded as a more precise version of the statistical syllogism:

> **prob**$(Bx/Ax)$ is high.
> $Ac$
> ________________
> Therefore, $Bc$.

Clearly, the conclusion of the statistical syllogism does not follow deductively from the premises. Furthermore, although the premises may often make it reasonable to accept the conclusion, that is not always the case. For instance, I may know that most ravens are black, and José is a raven, but I may also know that José is an albino raven and hence is not black. The premises of the statistical syllogism can at most create a presumption in favor of the conclusion, and that presumption can be defeated by contrary information. In other words, the inference licensed by this rule must be a defeasible inference. The inference is a reasonable one in the absence of conflicting information, but it is possible to have conflicting information in the face of which the inference becomes unreasonable.

In general, if $P$ is a defeasible reason for $Q$, there can be two kinds of defeaters for $P$. *Rebutting defeaters* are reasons for denying $Q$ in the face of $P$. To be contrasted with rebutting defeaters are *undercutting defeaters*, which attack the connection between the defeasible reason and its conclusion rather than attacking the conclusion itself. For example, something's looking red to me is a reason for me to think that it is red. But if I know that $x$ is illuminated by red lights and such illumination often makes things look red when they are not, then I cannot be justified in believing that $x$ is red on the basis of its looking red to me. What I know about the illumination constitutes an undercutting defeater for my defeasible reason. An undercutting defeater is a reason for denying that $P$ would not be true unless $Q$ were true. To illustrate, knowing about the peculiar illumination gives me a reason for denying that $x$ would not look red to me unless it were red. The technical details of reasoning with defeasible reasons and defeaters are discussed in my (1995). A simple sketch of the theory can be found in my (2007a).

As a first approximation, the statistical syllogism can be formulated as follows:

(3.1) If $r > .5$ then "$Ac$ and $\mathbf{prob}(Bx/Ax) \geq r$" is a defeasible reason for "$Bc$", the strength of the reason depending upon the value of $r$.

It is illuminating to consider how this rule handles the lottery paradox (Kyburg 1961). Suppose you hold one ticket in a fair lottery consisting of one million tickets, and suppose it is known that one and only one ticket will win. Observing that the probability is only .000001 of a ticket being drawn given that it is a ticket in the lottery, it seems reasonable to accept the conclusion that your ticket will not win. But by the same reasoning, it will be reasonable to believe, for each ticket, that it will not win. However, these conclusions conflict jointly with something else we are justified in believing, namely, that some ticket will win. Assuming that we cannot be justified in believing each member of an explicitly contradictory set of propositions, it follows that we are not warranted in believing of each ticket that it will not win. But this is no problem for our rule of statistical syllogism as long as it provides only a defeasible reason. What is happening in the lottery paradox is that the defeasible reason is defeated.

The lottery paradox is a case in which we have defeasible reasons for a number of conclusions but they collectively defeat one another. This illustrates the *principle of collective defeat*. This principle will turn out to be of considerable importance in probability theory, so I will say a bit more about it. Starting from propositions we are objectively justified in believing, we may be able to construct arguments supporting some further propositions. But that does not automatically make those further propositions warranted, because some propositions supported in that way may be defeaters for steps of some of the other arguments. That is what happens in cases of collective defeat. Suppose we are warranted in believing some proposition $R$ and we have equally good defeasible reasons for each of $P_1,\ldots,P_n$, where $\{P_1,\ldots,P_n\}$ is a minimal set of propositions deductively inconsistent with $R$ (that is, it is a set deductively inconsistent with $R$ and has no proper subset that is deductively inconsistent with $R$). Then for each $i$, the conjunction "$R$ & $P_1$ & ... & $P_{i-1}$ & $P_{i+1}$ & ... & $P_n$" entails $\sim P_i$. Thus by combining this entailment with the arguments for $R$ and $P_1,...,P_{i-1},P_{i+1},...,P_n$ we obtain an argument for $\sim P_i$ that is as good as the argument for $P_i$. Then we have equally strong support for both $P_i$ and $\sim P_i$, and hence we could not reasonably believe either on this basis, that is, neither is warranted. This holds for each $i$, so none of the $P_i$ is warranted. They collectively defeat one another. Thus the simplest version of the principle of collective defeat can be formulated as follows:

(3.2) If we are warranted in believing $R$ and we have equally good independent defeasible reasons for each member of a minimal set of propositions deductively inconsistent with $R$, and none of these defeasible reasons is defeated in any other way, then none of the propositions in the set is warranted on the basis of these defeasible reasons.

Although the principle of collective defeat allows the principle (3.1) of statistical syllogism to escape the lottery paradox, that principle is not adequate as it stands. There must be some constraints on what properties $A$ and $B$ can be employed in the statistical syllogism. At the very least, we need a constraint to rule out disjunctions. It turns out that disjunctions create repeated difficulties throughout the theory of probabilistic reasoning. This is easily illustrated for (3.1). For instance, it is a theorem of the probability calculus that $\mathbf{prob}(Fx/Gx \vee Hx) \geq \mathbf{prob}(Fx/Gx) \cdot \mathbf{prob}(Gx/Gx \vee Hx)$. Consequently, if $\mathbf{prob}(Fx/Gx)$ and $\mathbf{prob}(Gx/Gx \vee Hx)$ are sufficiently large, it follows that $\mathbf{prob}(Fx/Gx \vee Hx) \geq r$. For example, because the vast majority of birds can fly and because there are many more birds than giant sea tortoises, it follows that most things that are either birds or giant sea tortoises can fly. If Herman is a giant sea tortoise, (3.1) would give us a reason for thinking that Herman can fly, but notice that this is based simply on the fact that most birds can fly, which should be irrelevant to whether Herman can fly. This indicates that arbitrary disjunctions cannot be substituted for $B$ in (3.1).

Nor can arbitrary disjunctions be substituted for $A$ in (3.1). By the probability calculus, $\mathbf{prob}(Fx \vee Gx/Hx) \geq \mathbf{prob}(Fx/Hx)$. Therefore, if $\mathbf{prob}(Fx/Hx)$ is high, so is $\mathbf{prob}(Fx \vee Gx/Hx)$. Thus, because most birds can fly, it is also true that most birds can either fly or swim the English Channel. By (3.1), this should be a reason for thinking that a starling with a broken wing can swim the English Channel, but obviously it is not.

There must be restrictions on the properties $A$ and $B$ in (3.1). To have a convenient label, let us say that $B$ is *projectible with respect to* $A$ iff (3.1) holds. What we have seen is that projectibility is not closed under disjunction, that is, neither of the following hold:

If $C$ is projectible with respect to both $A$ and $B$, then $C$ is projectible with respect to $(A \vee B)$.

If $A$ and $B$ are both projectible with respect to $C$, then $(A \vee B)$ is projectible with respect to $C$.

On the other hand, it seems fairly clear that projectibility is closed under conjunction.

In formulating the principle of statistical syllogism, we must build in an explicit projectibility constraint:

(A1) If $F$ is projectible with respect to $G$ and $r > .5$, then "$Gc$ & $\mathbf{prob}(Fx/Gx) \geq r$" is a defeasible reason for believing "$Fc$", the strength of the reason depending upon the value of $r$.

Of course, if we define projectibility in terms of (3.1), (A1) becomes a mere tautology, but the intended interpretation of (A1) is that *there is* a relation of projectibility between properties, holding in important cases, such that "$Gc$ & $\mathbf{prob}(Fx/Gx) \geq r$" is a defeasible reason for "$Fc$" when $F$ is projectible with respect to $G$. To have a fully adequate theory we must augment (A1) with an account of projectibility, but that proves very difficult and I have no account to propose. At best, our conclusions about closure conditions provide a partial account. Because projectibility is closed under conjunction but not under disjunction, it follows that it is not closed under negation. Similar considerations establish that it is not closed under the formation of conditionals or biconditionals. It is not clear how it behaves with quantifiers. Although projectibility is not closed under negation, it seems likely that negations of "simple" projectible properties are projectible. A reasonable hypothesis is that there is a large class $\boldsymbol{P}$ containing most properties that are intuitively "logically simple", and whenever $A$ and $B$ are conjunctions of members of $\boldsymbol{P}$ and negations of members of $\boldsymbol{P}$, $A$ is projectible with respect to $B$. This is at best a sufficient condition for projectibility, however, because there are numerous cases of projectibility involving properties that are logically more complex than this.

The term "projectible" comes from the literature on induction. Goodman (1955) was the first to observe that principles of induction require a projectibility constraint.[2] I have deliberately chosen the term "projectible" in formulating the constraint on (A1). This is because in the theory of nomic probability, principles of induction become theorems rather than primitive postulates. The acceptance rules provide the epistemological machinery that make the theory run, and the projectibility constraint in induction turns out to derive from the projectibility constraint on (A1). It is the same notion of projectibility that is involved in both cases.

The reason provided by (A1) is only a defeasible reason. As with any defeasible reason, it can be defeated by having a reason for denying the conclusion. The reason for denying the conclusion constitutes a rebutting defeater. But there is also an important kind of undercutting defeater for (A1). In (A1), we infer the truth of $Fc$ on the basis of probabilities conditional on a limited set of facts about $c$ (the facts expressed by $Gc$). But if we know additional facts about $c$ that alter the probability, that defeats the defeasible

---

[2] For a useful compilation of articles on projectibility, see Stalker (1994).

reason:

(D1) If $F$ is projectible with respect to $H$ then "$Hc$ & **prob**($Fx/Gx$&$Hx$) ≠ **prob**($Fx/Gx$)" is an undercutting defeater for (A1).

I will refer to these as *subproperty defeaters*. (D1) amounts to a kind of "total evidence requirement". It requires us to make our inference on the basis of the most comprehensive facts regarding which we know the requisite probabilities.

(A1) is not the only defensible acceptance rule. There is another acceptance rule that is related to (A1) rather like modus tollens is related to modus ponens:

(A2) If $F$ is projectible with respect to $G$ and $r > .5$, then "~$Fc$ & **prob**($Fx/Gx$) ≥ $r$" is a defeasible reason for "~$Gc$", the strength of the reason depending upon the value of $r$.

(A2) is easily illustrated. On the basis of quantum mechanics, we can calculate that it is highly probable that an energetic electron will be deflected if it passes within a certain distance of a uranium atom. We observe that a particular electron was not deflected, and so conclude that it did not pass within the critical distance. Reasoning in this way with regard to the electrons used in a scattering experiment, we arrive at conclusions about the diameter of a uranium atom.

It seems clear that (A1) and (A2) are closely related. I suggest that they are consequences of a single stronger principle that, in chapter 7, is called "the statistical syllogism":

(SS) If $F$ is projectible with respect to $G$ and $r > .5$, then "**prob**($Fx/Gx$) ≥ $r$" is a defeasible reason for the conditional "$Gc \rightarrow Fc$", the strength of the reason depending upon the value of $r$.

(A1) can then be replaced by an instance of (SS) and modus ponens, and (A2) by an instance of (SS) and modus tollens. Accordingly, I will regard (SS) as the fundamental probabilistic acceptance rule. Just as in the case of (A1), when we use (SS) we are making an inference on the basis of a limited set of facts about $c$. That inference should be defeated if the probability can be changed by taking more facts into account. This indicates that the defeater for (A2) and (SS) should be the same as for (A1):

(D) If $F$ is projectible with respect to ($G$&$H$) then "$Hc$ & **prob**($Fx/Gx$&$Hx$) ≠ **prob**($Fx/Gx$)" is an undercutting defeater for (A1), (A2), and (SS).

To make (SS) work correctly, other defeaters are required as well. These are discussed at length in my (1990).[3]

I take it that (SS) is actually quite an intuitive acceptance rule. It amounts to a rule saying that, when $F$ is projectible with respect to $G$, if we know that most $G$'s are $F$, that gives us a reason for thinking of any particular object that it is an $G$ if it is a $F$. The only surprising feature of this rule is the projectibility constraint. (SS) is the basic epistemic principle from which all the rest of the theory of nomic probability is derived.

## 4. Direct Inference and Definite Probabilities

Nomic probability is a species of indefinite probability, but as I remarked above, for many purposes we are more interested in definite (single case) probabilities. In particular, definite probabilities are required for decision-theoretic purposes. The probabilities required for decision theory must have a strong epistemic element. For example, any decision on what odds to accept on a bet that Bluenose will win the next race must be based in part on what we know about Bluenose, and when our knowledge about Bluenose changes so will the odds we are willing to accept. Such probabilities are mixed physical/epistemic probabilities.

What I call *classical direct inference* aims to derive mixed physical/epistemic definite probabilities from indefinite probabilities. The basic idea behind classical direct inference was first articulated by Hans Reichenbach: in determining the probability that an individual $c$ has a property $F$, we find the narrowest reference class $X$ for which we have reliable statistics and then infer that $\mathbf{PROB}(Fc) = \mathbf{prob}(Fx/x \in X)$. For example, insurance rates are calculated in this way. There is almost universal agreement that direct inference is based upon some such principle as this, although there is little agreement about the precise form the theory should take. In my (1983) and (1984), I argued that classical direct inference should be regarded as proceeding in accordance with two epistemic rules. As formulated in chapter 7, the two rules are:

(CDI) If $F$ is projectible with respect to $G$, then "$\mathbf{prob}(Fx/Gx) = r$ & $Gc$ & $\mathbf{K}(P \leftrightarrow Fc)$" is a defeasible reason for "$\mathbf{PROB}(P) = r$".

(SCDI) If $F$ is projectible with respect to $H$ then "$\mathbf{K}Hc$ & $\mathbf{prob}(Fx/Gx\&Hx) \neq \mathbf{prob}(Fx/Gx)$" is an undercutting defeater for (CDI).

Principle (SCDI) formulates a kind of subproperty defeat for direct inference, because it says that probabilities based upon more specific information take precedence over those based upon less specific information. Note the projectibility constraint in these rules. That constraint is required to avoid various

---

[3] In my (1995) it is argued that these principles should be generalized further, but I will not pursue that here.

paradoxes of direct inference. Again, more defeaters are required to get a complete theory of direct inference.

To illustrate this account of direct inference, suppose we know that Herman is a 40-year-old resident of the United States who smokes. Suppose we also know that the probability of a 40-year-old resident of the United States having lung cancer is 0.1, but the probability of a 40-year-old smoker who resides in the United States having lung cancer is 0.3. If we know nothing else that is relevant we will infer that the probability of Herman having lung cancer is 0.3. (CDI) provides us with one defeasible reason for inferring that the probability is 0.1 and a second defeasible reason for inferring that the probability is 0.3. However, the latter defeasible reason is based upon more specific information, and so by (SCDI) it takes precedence, defeating the first defeasible reason and leaving us justified in inferring that the probability is 0.3.

I believe that (CDI) and (SCDI) are correct rules of classical direct inference, but I also believe that the nature of direct inference has been fundamentally misunderstood. Direct inference is taken to govern inferences from indefinite probabilities to definite probabilities, but it is my contention that such "classical" direct inference rests upon parallel inferences from indefinite probabilities to indefinite probabilities. The basic rule of classical direct inference is that if $F$ is projectible with respect to $G$ and we know "**prob**$(Fx/Gx) = r$ & **K**$(Gc)$" but do not know anything else about $c$ that is relevant, this gives us a reason to believe that PROB$(Fc) = r$. Typically, we will know $c$ to have other projectible properties $H$ but not know anything about the value of **prob**$(Fx/Gx\&Hx)$ and so be unable to use the latter in direct inference. But if the direct inference from "**prob**$(Fx/Gx) = r$" to "PROB$(Fc) = r$" is to be reasonable, there must be a presumption to the effect that **prob**$(Fx/Gx\&Hx) = r$. If there were no such presumption then we would have to regard it as virtually certain that **prob**$(Fx/Gx\&Hx) \neq r$ (after all, there are infinitely many possible values that **prob**$(Fx/Gx\&Hx)$ could have), and so virtually certain that there is a true subproperty defeater for the direct inference. This would make the direct inference to "PROB$(Fc) = r$" unreasonable. Thus classical direct inference presupposes the following principle regarding indefinite probabilities:

(DI) If $F$ is projectible with respect to $G$ then "**prob**$(Fx/Gx) = r$" is a defeasible reason for "**prob**$(Fx/Gx\&Hx) = r$".

Inferences in accord with (DI) comprise *non-classical direct inference*. (DI) amounts to a kind of principle of insufficient reason, telling us that if we have no reason to think otherwise, it is reasonable for us to anticipate that conjoining $H$ to $G$ will not affect the probability of $F$.

Nonclassical direct inference is related to another issue that is continually bothersome to probability practitioners. The probabilities PROB$(P)$ and PROB$(Q)$ are *independent* iff PROB$(P\&Q)$ = PROB$(P)\cdot$PROB$(Q)$. In the practical use of probabilities it is almost universally assumed, often apologetically, that probabilities are independent unless we have some reason for thinking otherwise. In most real-world applications of probabilities, if we did not make such assumptions about independence we would not be able to compute any of the complex probabilities that interest us. For example, suppose we know that the probability of a Chevrolet having a defective steering mechanism if it has power steering and was manufactured in a certain plant is .3, whereas the probability of its having a defective steering mechanism otherwise is only .01. We also know that the probability of a Chevrolet being manufactured in that plant is .33, and the probability of a Chevrolet having power steering is .85. If we know nothing else of relevance, we will normally assume that whether the car has power steering is independent of whether it was manufactured in that plant, and so compute

**PROB**(power-steering & plant) = .33 ·.85 = .28.

Then we can compute:

**PROB**(defect) = .3 ·.28 + .01·(1-.28) = .09.

We could not perform this, or similar computations, without a defeasible assumption of independence. This is precise what (DI) gives us. By the probability calculus,

$$\mathbf{prob}(Fx\&Hx/Gx) = \mathbf{prob}(Fx/Gx\&Hx)\cdot\mathbf{prob}(Hx/Gx).$$

Thus

$$\mathbf{prob}(Fx\&Hx/Gx) = \mathbf{prob}(Fx/Gx)\cdot\mathbf{prob}(Hx/Gx)$$

iff

$$\mathbf{prob}(Fx/Gx\&Hx) = \mathbf{prob}(Fx/Gx).$$

And the latter is what we get from (DI). So (DI) is equivalent to a defeasible assumption of statistical independence.

Although (DI) is a plausible principle, it is not self-evident. The only defense I have given for it so far is that it is required for the legitimacy of inferences we commonly make. That is a reason for thinking that it is true, but we would like to know why it is true. The answer to this question is supplied by the principle of agreement and the acceptance rule (A1). We have the following instance of (A1):

(4.1) If "$\mathbf{prob}(Fx/Xx) \approx_\delta r$" is projectible with respect to "$X \preceq G$" then "$H \preceq G$ & $\mathbf{prob}\big(\mathbf{prob}(F/X) \approx_\delta r \,/\, X \preceq G\big) = 1$" is a defeasible reason for "$\mathbf{prob}(Fx/Hx) \approx_\delta r$".

If we assume that the property "$\mathbf{prob}(Fx/Xx) \approx_\delta r$" is projectible with respect to "$X \preceq G$" whenever $F$ is projectible with respect to $G$, then it follows that:

(4.2) If $F$ is projectible with respect to $G$ then "$H \preceq G$ & $\mathbf{prob}\big(\mathbf{prob}(Fx/Xx) \approx_\delta r \ / \ X \preceq G\big) = 1$" is a defeasible reason for "$\mathbf{prob}(Fx/Hx) \approx_\delta r$".

By the principle of agreement, for each $\delta > 0$, "$\mathbf{prob}(Fx/Gx) = r$" entails "$\mathbf{prob}(\mathbf{prob}(Fx/Xx) \approx_\delta r \ / \ X \preceq G) = 1$", so it follows that:

(4.3) If $F$ is projectible with respect to $G$ then for each $\delta > 0$, "$H \preceq G$ & $\mathbf{prob}(Fx/Gx) = r$" is a defeasible reason for "$\mathbf{prob}(Fx/Hx) \approx_\delta r$".

I showed in my (1984) that (DI) follows from this.

Similar reasoning enables us to derive the following defeater for (DI):

(SDI) If $F$ is projectible with respect to $J$ and $\Box(\forall x)[(Gx\&Hx) \rightarrow Jx]$ then "$\mathbf{prob}(Fx/Gx) \neq \mathbf{prob}(Fx/Gx\&Jx)$" is an undercutting defeater for (DI).

I will refer to these as *subproperty defeaters* for nonclassical direct inference. Again, more defeaters are required to make non-classical direct inference work properly. See my (1990) for a discussion of this.

We now have two kinds of direct inference—classical and non-classical. Direct inference has traditionally been identified with classical direct inference, but I believe that it is most fundamentally non-classical direct inference. The details of classical direct inference are all reflected in non-classical direct inference. If we could identify definite probabilities with certain indefinite probabilities, we could derive the theory of classical direct inference from the theory of non-classical direct inference. This can be done as follows. Let $\mathbb{K}$ be the conjunction of all the propositions the cognizer knows to be true. For any proposition $P$, let *P-worlds* be physically possible worlds at which $P$ is true. Then using the same proportion function that was employed in the definition of nomic probabilities, we can define

$\mathrm{PROB}(P)$ is the proportion of $\mathbb{K}$-worlds that are also $P$-worlds.

If $Q$ is consistent with $\mathbb{K}$ then $\mathrm{PROB}(P/Q)$ is the proportion of $(Q\&\mathbb{K})$-worlds that are also $P$-worlds.

This is a natural way of understanding definite probabilities, and the theory of nomic probabilities then enables us to prove the following representation theorem:

$\mathrm{PROB}(Fa) = \mathbf{prob}(Fx \ / \ x = a \ \& \ \mathbb{K})$.

This is the probability of something having the property $F$ given that it is the object $a$ and given everything we know. Technically, this is an indefinite probability, but it is only about the object $a$. Presumably, this nomic probability can only be evaluated by using non-classical direct inference, and the structure of that reasoning mirrors the structure of the reasoning involved intuitively in classical direct inference. More generally, we can prove:

> If it is physically necessary that $[\mathbf{K} \rightarrow (Q \leftrightarrow Sa_1...a_n)]$ and that $[(Q\&\mathbf{K}) \rightarrow (P \leftrightarrow Ra_1...a_n)]$, and $Q$ is consistent with $\mathbf{K}$, then $\text{PROB}(P/Q) = \mathbf{prob}(Rx_1...x_n \,/\, Sx_1...x_n \;\&\; x_1 = a_1 \;\&\; ... \;\&\; x_n = a_n \;\&\; \mathbf{K})$.

This makes precise the way in which the definite probabilities at which we arrive by direct inference are mixed physical/epistemic probabilities. We obtain definite probabilities by considering indefinite probabilities conditional on properties we know the objects in question to possess.

Given this reduction of definite probabilities to indefinite probabilities, it we can *derive* principles (CDI) and (SCDI) of classical direct inference from our principles of non-classical direct inference, and hence indirectly from (SS) and the calculus of nomic probabilities. The upshot of all this is that the theory of direct inference, both classical and nonclassical, consists of a sequence of *theorems* in the theory of nomic probability. We require no new assumptions in order to get direct inference. At the same time, we have made clear sense of mixed physical/epistemic probabilities.

## 5. Indefinite Probabilities and Probability Distributions

It is a disturbing fact that contemporary philosophers and probability theorists often seem completely insensitive to the distinction between definite and indefinite probabilities. My experience has been that as soon as one sees the distinction, it seems obvious and important. Why are people oblivious to it? After all, prior to subjective probability becoming popular, most theories of probability were frequency-based and were accordingly theories of indefinite probability.

Part of the explanation for this strange conceptual blindness can be traced to the popularity of subjective probability. Subjective probability has no way of accommodating indefinite probabilities. Because subjective probabilities are degrees of belief, they are inherently definite probabilities. Still, everyone knows that there are theories of objective probability, so why do they entirely overlook indefinite probabilities?

I think a large part of the explanation turns on confusing talk of indefinite probabilities with talk of probability distributions. Where $x$ is a variable ranging over the members of a set $X$ of "items" (objects, measurements, or whatever you like) and $Fx$ is a formula with free variable $x$, the probability distribution for $Fx$ over $X$ is written as "prob($Fx$)" (with "random variable $x$"), is the set of all probabilities prob($Fc$) for different members $c$ of $X$. That is, it is $\{\text{prob}(Fc) \mid c \in X\}$. "prob" here can be any kind of definite probability we like. We can also talk about the conditional probability distribution prob($Fx/Gx$), which is $\{\text{prob}(Fc/Gc) \mid c \in X\}$. Because probability distributions are written using "random variables", they look misleadingly like indefinite probabilities. But they are not—they are sets of definite probabilities. For instance, if we take prob to be **PROB**, the probability distribution over human beings for "*x* will get lung cancer" conditional on "*x* smokes" is the set of all probabilities **PROB**(*c* will get lung cancer/*c* smokes) for each human being *c*. It is a *set* of numbers, not a single number. By contrast, the indefinite probability **prob**(*x* will get lung cancer/*x* smokes & *x* is human) is a single number.

Given the probability distribution of *Fx* over the set *X*, the *distribution function* is said to be the function that assigns a probability to "a randomly chosen member of *X* that satisfies *Fx*" and "a randomly chosen member of *X* that does not satisfy *Fx*".[4] There is a temptation to think that by talking about the probability of a randomly chosen member of *X* satisfying *Fx*, we can get the effect of indefinite probabilities by talking just about definite probabilities. However, we can do so only by smuggling in indefinite probabilities from the start. The expression "the probability of a randomly chosen member of *X* satisfying *Fx*" is ambiguous. It could mean "the probability of an arbitrary member of *X* satisfying *Fx*"—the indefinite probability—or it could mean "where *c* is a member of *X* that was chosen randomly, **PROB**(*Fc*)". The latter does not smuggle in indefinite probabilities, but the value need not be the same as the indefinite probability. Suppose *c* is the randomly chosen member of *X*. To say that it was chosen randomly is just to talk about the method of selection. It tells us nothing about what knowledge we have of *c*. If, in fact, we know nothing about *c* except that it is a member of *X*, then by direct inference, **PROB**(*Fc*) will be the same as the indefinite probability. But suppose *X* is a set of humans, and for each member of the set, their gender is obvious. Suppose the indefinite probability of a member of *X* being male is 1/2. Nevertheless, suppose it is obvious to us that *c* is female. Then, despite the fact that *c* was chosen at random, **PROB**(*c* is male) = 0.

I think the right way to understand distribution functions is in terms of indefinite probabilities, and that works fine for the purposes of mathematical probability theory. But anyone who thinks that definite probabilities are the fundamental kind of probability (for example, a subjectivist) cannot do that. He is thus precluded from using probability distributions to do the work of indefinite probabilities.

## 6. Induction

The values of some nomic probabilities are derivable from the values of others using the calculus of nomic probabilities or the theory of nonclassical direct inference, but our initial knowledge of nomic probabilities must result from empirical observation of the world around us. This is accomplished by

---

[4] This is to talk about distribution functions for two-valued random variables. In general, things can get more complicated.

*statistical induction*. We observe the relative frequency of $F$'s in a sample of $G$'s, and then infer that **prob**$(Fx/Gx)$ is approximately equal to that relative frequency. One of the main strengths of the theory of nomic probability is that precise principles of induction can be derived from (and hence justified on the basis of) the acceptance rules and computational principles we have already endorsed. This leads to a solution of sorts to the problem of induction.

Principles of statistical induction are principles telling us how to estimate probabilities on the basis of observed relative frequencies in finite samples. The problem of constructing such principles is sometimes called *the problem of inverse inference*. All theories of inverse inference are similar in certain respects. In particular, they all make use of some form of Bernoulli's theorem. In its standard formulation, Bernoulli's theorem tells us that if we have $n$ objects $b_1,...,b_n$ and for each $i$, $\text{PROB}(Ab_i) = p$, and any of these objects being $A$ is statistically independent of which others of them are $A$, then the probability is high that the relative frequency of $A$'s among $b_1,...,b_n$ is approximately $p$. Furthermore, the probability increases and the degree of approximation improves as $n$ is made larger. These probabilities can be computed by noting that on the stated assumption of independence, it follows from the probability calculus that

$$\text{PROB}(Ab_1 \,\&\, ... \,\&\, Ab_r \,\&\, {\sim}Ab_{r+1} \,\&\, ... \,\&\, {\sim}Ab_n)$$

$$= \text{PROB}(Ab_1)\cdot...\cdot\text{PROB}(Ab_r)\cdot\text{PROB}({\sim}Ab_{r+1})\cdot...\cdot\text{PROB}({\sim}Ab_n)$$

$$= p^r(1-p)^{n-r}.$$

There are $n!/r!(n-r)!$ distinct ways of assigning $A$-hood among $b_1,...,b_n$ such that freq$[A/\{b_1,...,b_n\}] = r/n$, so it follows by the probability calculus that

$$\text{PROB}(\text{freq}[A/\{b_1,...,b_n\}] = r/n) = n!p^r(1-p)^{n-r}/r!(n-r)!$$

The right side of this equation is the formula for the binomial distribution. An interval $[p-\varepsilon,p+\varepsilon]$ around $p$ will contain just finitely many fractions $r/n$ with denominator $n$, so we can calculate the probability that the relative frequency has any one of those values, and then the probability of the relative frequency being in the interval is the sum of those probabilities.

Thus far everything is uncontroversial. The problem is what to do with the probabilities resulting from Bernoulli's theorem. Most theories of inverse inference, including most of the theories embodied in contemporary statistical theory, can be regarded as variants of a single intuitive argument that goes as follows. Suppose **prob**$(Ax/Bx) = p$, and all we know about $b_1,...,b_n$ is that they are $B$'s. Then by classical direct inference we can infer that for each $i$, $\text{PROB}(Ab_i) = p$. If the $b_i$'s seem intuitively unrelated to one another then it seems reasonable to suppose they are statistically independent and so we can use Bernoulli's theorem and conclude that it is extremely probable that the observed relative frequency $r/n$ lies in a small interval $[p-\varepsilon,p+\varepsilon]$ around $p$. This entails conversely that $p$ is within $\varepsilon$ of $r/n$, that is, $p$ is in the interval

$[(r/n)-\varepsilon,(r/n)+\varepsilon]$. This becomes our estimate of $p$.

A problem for the intuitive argument concerns what to do with the conclusion that it is very probable that the observed frequency is within $\varepsilon$ of $p$. It is tempting to suppose we can use the statistical syllogism (SS) and reason:

> If **prob**$(Ax/Bx) = p$ then **PROB**(freq$[A/\{b_1,\ldots,b_n\}] \in [p-\varepsilon,p+\varepsilon]$) is approximately equal to 1,

so,

> if **prob**$(Ax/Bx) = p$ then freq$[A/\{b_1,\ldots,b_n\}] \in [p-\varepsilon,p+\varepsilon]$.

The latter entails

> If freq$[A/\{b_1,\ldots,b_n\}] = r/n$ then **prob**$(Ax/Bx) \in [(r/n)-\varepsilon,(r/n)+\varepsilon]$.

A rather shallow difficulty for this reasoning is that it is an incorrect use of (SS). (SS) concerns indefinite probabilities, while Bernoulli's theorem supplies us with definite probabilities. But let us waive that difficulty for the moment, because there is a much more profound difficulty. This is that the probabilities we obtain in this way have the structure of the lottery paradox. Given any point $q$ in the interval $[p-\varepsilon,p+\varepsilon]$, we can find a small number $\lambda$ around it such that if we let $\mathbf{I}_q$ be the union of two intervals $[0,q-\lambda]\cup[q+\lambda,1]$, the probability of freq$[A/\{b_1,\ldots,b_n\}]$ being in $\mathbf{I}_q$ is as great as the probability of its being in $[p-\varepsilon,p+\varepsilon]$. This is diagrammed in figure A.1. The probability of the frequency falling in any interval is represented by the area under the curve corresponding to that interval. The curve is reflected about the $x$ axis so that the probability for the interval $[p-\varepsilon,p+\varepsilon]$ can be represented by the shaded area above the axis and the probability for the interval $\mathbf{I}_q$ represented by the shaded area below the axis.

Next notice that we can construct a finite set $q_1,\ldots,q_k$ of points in $[p-\varepsilon,p+\varepsilon]$ such that the "gaps" in the $\mathbf{I}_{q_i}$ collectively cover $[p-\varepsilon,p+\varepsilon]$. This is diagrammed in figure A.2. For each $i \leq k$, we have as good a reason for believing that freq$[A/\{b_1,\ldots,b_n\}]$ is in $\mathbf{I}_{q_i}$ as we do for thinking it is in $[p-\varepsilon,p+\varepsilon]$, but these conclusions are jointly inconsistent. This is analogous to the lottery paradox. We have a case of collective defeat and thus are unjustified in concluding that the relative frequency is in the interval $[p-\varepsilon,p+\varepsilon]$.

The intuitive response to the "lottery objection" consists of noting that $[p-\varepsilon,p+\varepsilon]$ is an interval while the $\mathbf{I}_q$ are not. Somehow, it seems right to make an inference regarding intervals when it is not right to make the analogous inference regarding "gappy" sets. That is the line taken in orthodox statistical inference when confidence intervals are employed. But it is very hard to see why this should be the case, and some heavy duty argument is needed here to justify the whole procedure.

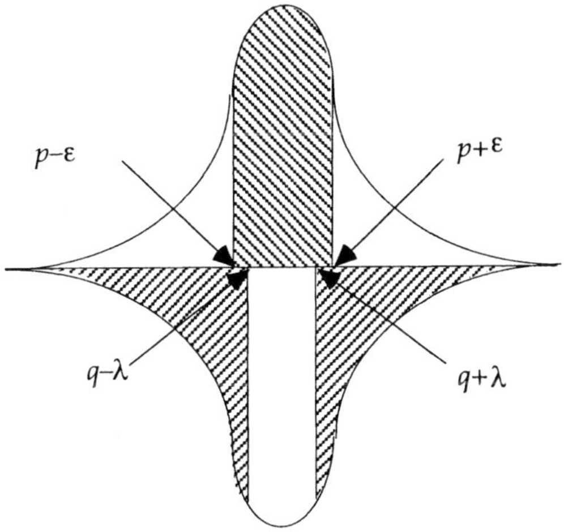


Figure A.1. Intervals on the tails with the same probability as an interval at the maximum

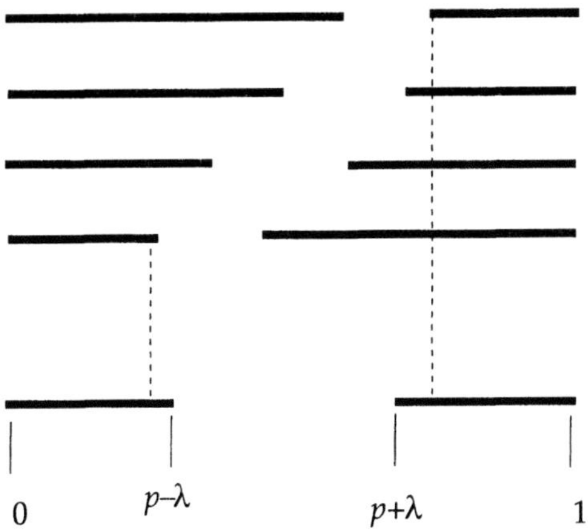


Figure A.2. A collection of small gaps covers a large gap

In sum, when we try to make the intuitive argument precise, it becomes apparent that it contains major gaps. This does not constitute an utter condemnation of the intuitive argument. Because it is so intuitive, it would be surprising if it were not at least approximately right. Existing statistical theories are mathematically sophisticated, but from an epistemological perspective they tend to be *ad hoc* jury-rigged affairs without adequate foundations. Still, it seems likely that there are some sound intuitions that statisticians are trying to capture with these theories. The problem is to turn the intuitive argument into a rigorous and defensible argument. That, in effect, is what my account of statistical induction does. The argument will undergo three kinds of repairs, creating what I call *the statistical induction argument*. First, it will be reformulated in terms of indefinite probabilities, thus enabling us to make legitimate use of (SS). Second, it will be shown that the undefended assumption of statistical independence can be filled by nonclassical direct inference. Third, the final step of the argument will be scrapped and replaced by a more complex argument not subject to the lottery paradox. This more complex argument will employ a principle akin to the *Likelihood Principle* of classical statistical inference.

The details of the statistical induction argument are complicated, and can be found in full in my (1990). I will try to convey the gist of the argument by focusing on a special case. Normally, **prob**$(Ax/Bx)$ can have any value from 0 to 1. The argument is complicated by the fact that there are infinitely many possible values. Let us suppose instead that we somehow know that **prob**$(Ax/Bx)$ has one of a finite set of values $p_1,\ldots,p_k$. If we have observed a sample $X = \{b_1,\ldots,b_n\}$ of $B$'s and noted that only $b_1,\ldots,b_r$ are $A$'s (where $A$ and $\sim A$ are projectible with respect to $B$), then the relative frequency freq$[A/X]$ of $A$'s in $X$ is $r/n$. From this we want to infer that **prob**$(Ax/Bx)$ is approximately $r/n$. Our reasoning proceeds in two stages, the first stage employing the theory of nonclassical direct inference, and the second stage employing the statistical syllogism.

*Stage I*

By "$x_1,\ldots,x_n$ are distinct" I mean that for each $i$ and $j$, if $i \neq j$ then $x_i \neq x_j$. Let us abbreviate "$x_1,\ldots,x_n$ are distinct & $Bx_1$ & ... & $Bx_n$ & **prob**$(Ax/Bx) = p$" as "$\theta_p$". When $r \leq n$, we have by the probability calculus:

(6.1)
$$\begin{aligned}&\mathbf{prob}(Ax_1 \,\&\, \ldots \,\&\, Ax_r \,\&\, {\sim}Ax_{r+1} \,\&\, \ldots \,\&\, {\sim}Ax_n \,/\, \theta_p)\\ &= \mathbf{prob}(Ax_1 \,/\, Ax_2 \,\&\, \ldots \,\&\, Ax_r \,\&\, {\sim}Ax_{r+1} \,\&\, \ldots \,\&\, {\sim}Ax_n \,\&\, \theta_p)\\ &\cdot \ldots \cdot \mathbf{prob}(Ax_r \,/\, {\sim}Ax_{r+1} \,\&\, \ldots \,\&\, {\sim}Ax_n \,\&\, \theta_p)\\ &\cdot \mathbf{prob}({\sim}Ax_{r+1} \,/\, {\sim}Ax_{r+2} \,\&\, \ldots \,\&\, {\sim}Ax_n \,\&\, \theta_p) \cdot \ldots \cdot \mathbf{prob}({\sim}Ax_n \,/\, \theta_p).\end{aligned}$$

Making $\theta_p$ explicit:

(6.2)
$$\begin{aligned}&\mathbf{prob}(Ax_i \,/\, Ax_{i+1} \,\&\, \ldots \,\&\, Ax_r \,\&\, {\sim}Ax_{r+1} \,\&\, \ldots \,\&\, {\sim}Ax_n \,\&\, \theta_p)\\ &= \mathbf{prob}\big(Ax_i \,/\, x_1,\ldots,x_n \text{ are distinct} \,\&\, Bx_1 \,\&\, \ldots \,\&\, Bx_n \,\&\, Ax_{i+1} \,\&\, \ldots \,\&\, Ax_r\\ &\qquad \&\, {\sim}Ax_{r+1} \,\&\, \ldots \,\&\, {\sim}Ax_n \,\&\, \mathbf{prob}(Ax/Bx) = p\big).\end{aligned}$$

Projectibility is closed under conjunction, so "$Ax_i$" is projectible with respect to "$Bx_1$ & ... & $Bx_n$ & $x_1,...,x_n$ are distinct & $Ax_{i+1}$ & ... & $Ax_r$ & ~$Ax_{r+1}$ & ... & ~$Ax_n$". Given principles we have already endorsed it can be proven that whenever "$Ax_i$" is projectible with respect to "$Fx$", it is also projectible with respect to "$Fx$ & **prob**$(Ax/Bx) = p$". Consequently, "$Ax_i$" is projectible with respect to the reference property of (6.2). Thus non-classical direct inference gives us a reason for believing that

$$\textbf{prob}(Ax_i \ / \ Ax_{i+1} \ \& \ ... \ \& \ Ax_r \ \& \ {\sim}Ax_{r+1} \ \& \ ... \ \& \ {\sim}Ax_n \ \& \ \theta_p)$$

$$= \textbf{prob}\big(Ax_i \ / \ Bx_i \ \& \ \textbf{prob}(Ax/Bx) = p\big),$$

which by principle (PPROB) equals $p$. Similarly, non-classical direct inference gives us a reason for believing that if $r < i \leq n$ then

(6.3) $\textbf{prob}({\sim}Ax_i \ / \ {\sim}Ax_{i+1} \ \& \ ... \ \& \ {\sim}Ax_n \ \& \ \theta_p) = 1-p.$

Then from (6.1) we have:

(6.4) $\textbf{prob}(Ax_1 \ \& \ ... \ \& \ Ax_r \ \& \ {\sim}Ax_{r+1} \ \& \ ... \ \& \ {\sim}Ax_n \ / \ \theta_p) = p^r(1-p)^{n-r}.$

"freq[A / $\{x_1,...,x_n\}$] = $r/n$" is equivalent to a disjunction of $n!/r!(n-r)!$ pairwise incompatible disjuncts of the form "$Ax_1$ & ... & $Ax_r$ & ~$Ax_{r+1}$ & ... & ~$Ax_n$", so by the probability calculus:

(6.5) $\textbf{prob}\big(\text{freq}[A \ / \ \{x_1,...,x_n\}] = r/n \ / \ \theta_p\big) = n!p^r(1-p)^{n-r}/r!(n-r)!$

This is the formula for the binomial distribution.

This completes stage I of the statistical induction argument. This stage reconstructs the first half of the intuitive argument described above. Note that it differs from that argument in that it proceeds in terms of indefinite probabilities rather than definite probabilities, and by using nonclassical direct inference it avoids having to make unwarranted assumptions about independence. In effect, nonclassical direct inference gives us a reason for expecting independence unless we have evidence to the contrary. All of this is a consequence of the statistical syllogism and the calculus of nomic probabilities.

*Stage II*

The second half of the intuitive argument ran afoul of the lottery paradox and seems to me to be irreparable. I propose to replace it with an argument using (A2). I assume at this point that if $A$ is a projectible property then (for variable $X$) "freq[$A/X$] $\neq r/n$" is a projectible property. Thus the following conditional probability, derived from (6.5), satisfies the projectibility constraint of our acceptance rules:

(6.6) $\textbf{prob}\big(\text{freq}[A \ / \ \{x_1,...,x_n\}] \neq r/n \ / \ \theta_p\big) = 1 - n!p^r(1-p)^{n-r}/r!(n-r)!$

Let $\mathbf{b}(n,r,p) = n!p^r(1-p)^{n-r}/r!(n-r)!$ . For sizable $n$, $\mathbf{b}(n,r,p)$ is almost always quite small. For example, $\mathbf{b}(50,20,.5) = .04$. Thus by (A2) and (6.6), for each choice of $p$ we have a defeasible reason for believing that $\langle b_1,\ldots,b_n\rangle$ does not satisfy $\theta_p$, that is, for believing

$$\sim(b_1,\ldots,b_n \text{ are distinct } \& \ Bb_1 \ \& \ \ldots \ \& \ Bb_n \ \& \ \mathbf{prob}(Ax/Bx) = p).$$

As we know that "$b_1,\ldots,b_n$ are distinct & $Bb_1$ & ... & $Bb_n$" is true, this gives us a defeasible reason for believing that $\mathbf{prob}(Ax/Bx) \neq p$. But we know that for some one of $p_1,\ldots,p_k$, $\mathbf{prob}(Ax/Bx) = p_i$. This case is much like the case of the lottery paradox. For each $i$ we have a defeasible reason for believing that $\mathbf{prob}(Ax/Bx) \neq p_i$, but we also have a counterargument for the conclusion that $\mathbf{prob}(Ax/Bx) = p_i$, namely:

$\mathbf{prob}(Ax/Bx) \neq p_1$

$\mathbf{prob}(Ax/Bx) \neq p_2$

.
.

$\mathbf{prob}(Ax/Bx) \neq p_{i-1}$

$\mathbf{prob}(Ax/Bx) \neq p_{i+1}$

.
.

$\mathbf{prob}(Ax/Bx) \neq p_k$

For some $j$ between 1 and $k$, $\mathbf{prob}(Ax/Bx) = p_j$.

Therefore, $\mathbf{prob}(Ax/Bx) = p_i$.

However, there is an important difference between this case and a fair lottery. For each $i$, we have a defeasible reason for believing that $\mathbf{prob}(Ax/Bx) \neq p_i$, but these reasons are not all of the same strength because the probabilities assigned by (6.6) differ for the different $p_i$'s. The counterargument is only as good as its weakest link, so for some of the $p_i$'s, the counterargument may not be strong enough to defeat the defeasible reason for believing that $\mathbf{prob}(Ax/Bx) \neq p_i$. This will result in there being a subset $\mathbf{R}$ (the *rejection class*) of $\{p_1,\ldots,p_k\}$ such that we can conclude that that for each $p \in \mathbf{R}$, $\mathbf{prob}(Ax/Bx) \neq p$, and hence $\mathbf{prob}(Ax/Bx) \notin \mathbf{R}$. Let $\mathbf{A}$ (the *acceptance class*) be $\{p_1,\ldots,p_k\} - \mathbf{R}$. It follows that we are justified in believing that $\mathbf{prob}(Ax/Bx) \in \mathbf{A}$. $\mathbf{A}$ will consist of those $p_i$'s closest in value to $r/n$. Thus we can think of $\mathbf{A}$ as an interval around the observed frequency, and we are justified in believing that $\mathbf{prob}(Ax/Bx)$ lies in that interval.

To fill in some details, let us abbreviate "$r/n$" as "$f$", and make the simplifying assumption that for some $i$, $p_i = f$. $\mathbf{b}(n,r,p)$ will always be highest for this value of $p$, which means that (6.6) provides us with a weaker reason for believing that $\mathbf{prob}(Ax/Bx) \neq f$ than it does for believing that $\mathbf{prob}(Ax/Bx) \neq p_j$ for any of the other $p_j$'s. It follows that $f$ cannot be in the rejection class, because each step of the counterargument is better than the reason for believing that **prob**$(Ax/Bx) \neq f$. On the other hand, "**prob**$(Ax/Bx) \neq f$" will be the weakest step of the counterargument for every other $p_j$. Thus what determines whether $p_j$ is in the rejection class is simply the comparison of $\mathbf{b}(n,r,p_j)$ to $\mathbf{b}(n,r,f)$. A convenient way to encode this comparison is by considering the ratio

(6.7) $L(n,r,p) = \mathbf{b}(n,r,p)/\mathbf{b}(n,r,f) = (p/f)^{nf} \cdot ((1-p)/1-f)^{n(1-f)}$.

$L(n,r,p)$ is the *likelihood ratio* of "**prob**$(Ax/Bx) = p$" to "**prob**$(Ax/Bx) = f$". The smaller the likelihood ratio, the stronger is our on-balance reason for believing (despite the counterargument) that **prob**$(Ax/Bx) \neq p$, and hence the more justified we are in believing that **prob**$(Ax/Bx) \neq p$. Each degree of justification corresponds to a minimal likelihood ratio, so we can take the likelihood ratio to be a measure of the degree of justification. For each likelihood ratio $\alpha$ we obtain the *$\alpha$-rejection class* $\mathbf{R}_\alpha$ and the *$\alpha$-acceptance class* $\mathbf{A}_\alpha$:

(6.8) $\mathbf{R}_\alpha = \{p_i \mid L(n,r,p_i) \leq \alpha\}$.

(6.9) $\mathbf{A}_\alpha = \{p_i \mid L(n,r,p_i) > \alpha\}$.

We are justified to degree $\alpha$ in rejecting the members of $\mathbf{R}_\alpha$, and hence we are justified to degree $\alpha$ in believing that **prob**$(Ax/Bx)$ is a member of $\mathbf{A}_\alpha$. If we plot the likelihood ratios, we get a bell curve centered around $r/n$, with the result that $\mathbf{A}_\alpha$ is an interval around $r/n$ and $\mathbf{R}_\alpha$ consists of the tails of the bell curve. This is shown in figure A.3. In interpreting this curve, remember that low likelihood ratios correspond to a high degree of justification for *rejecting* that value for **prob**$(Ax/Bx)$, and so the region around $r/n$ consists of those values we cannot reject. That is, it consists of those values that might be the actual value.[5]

I have been discussing the idealized case in which we know that **prob**$(Ax/Bx)$ has one of a finite set of values, but it is shown in my (1990) that the argument can be generalized to apply to the continuous case as well. The argument provides us with justification for believing that **prob**$(Ax/Bx)$ lies in a precisely defined interval around the observed relative frequency, the width of the interval being a function of the degree of justification. For illustration, some typical values of the acceptance interval are listed in table A.1.

The statistical induction argument bears obvious similarities to orthodox statistical reasoning. It would be very surprising if that were not so, because all theories of inverse inference are based upon variants of the intuitive argument discussed above. But it should also be emphasized that if you try to list the exact points of similarity between the statistical induction argument and orthodox statistical reasoning concerning confidence intervals or significance testing, there are as many important differences as there are similarities.

---

[5] Making this reasoning precise requires a theory of defeasible reasoning that accommodates "diminishers". See my (2002a).

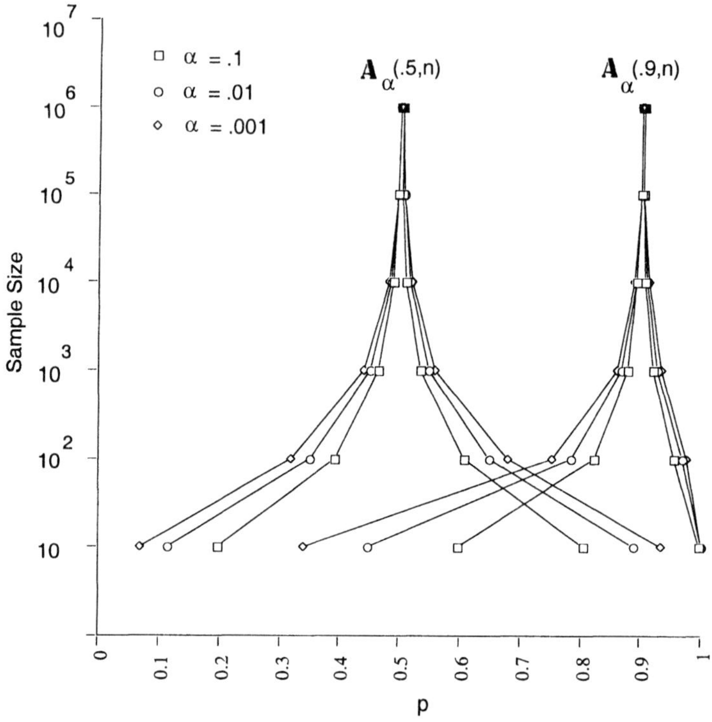


Figure A.3. Acceptance intervals

$\mathbf{A}_\alpha(.5,n)$

| α | $n$ = 10 | $10^2$ | $10^3$ | $10^4$ | $10^5$ | $10^6$ |
|---|---|---|---|---|---|---|
| .1 | [.196,.804] | [.393,.607] | [.466,.534] | [.489,.511] | [.496,.504] | [.498,.502] |
| .01 | [.112,.888] | [.351,.649] | [.452,.548] | [.484,.516] | [.495,.505] | [.498,.502] |
| .001 | [.068,.932] | [.320,.680] | [.441,.559] | [.481,.519] | [.494,.506] | [.498,.502] |

$\mathbf{A}_\alpha(.9,n)$

| α | $n$ = 10 | $10^2$ | $10^3$ | $10^4$ | $10^5$ | $10^6$ |
|---|---|---|---|---|---|---|
| .1 | [.596,.996] | [.823,.953] | [.878,.919] | [.893,.907] | [.897,.903] | [.899,.901] |
| .01 | [.446,1.00] | [.785,.967] | [.868,.927] | [.890,.909] | [.897,.903] | [.899,.901] |
| .001 | [.338,1.00] | [.754,.976] | [.861,.932] | [.888,.911] | [.897,.903] | [.899,.901] |

Table A.1. Values of $\mathbf{A}_\alpha(f,n)$

These include the use of indefinite probabilities in place of definite probabilities, the use of nonclassical direct inference in place of assumptions of independence, the *derivation* of the principle governing the use of likelihood ratios rather than the simple postulation of the likelihood principle, and the projectibility constraint. The latter deserves particular emphasis. A few writers have suggested general theories of statistical inference that are based on likelihood ratios and agree closely with the present account when applied to the special case of statistical induction. However, such theories have been based merely on statistical intuition, without an underlying rationale of the sort given here, and they are all subject to counterexamples having to do with projectibility.

Enumerative induction differs from statistical induction in that our inductive sample must consist of *B*'s *all of which are A*'s and our conclusion is the generalization "$Bx \Rightarrow Ax$". It is shown in my (1984a) and (1990) that in the special case in which the observed relative frequency is 1, it is possible to extend the statistical induction argument to obtain an argument for the conclusion "$Bx \Rightarrow Ax$". Thus enumerative induction can also be justified on the basis of nomic probability.

## 7. Conclusions

To briefly recapitulate, the theory of nomic probability aims to make objective probability philosophically respectable by providing a mathematically precise theory of probabilistic reasoning. This theory has two primitive parts: (1) a set of computational principles comprising a strengthened probability calculus, and (2) an acceptance rule licensing defeasible inferences from high probabilities. Some of the computational principles are novel, but the resulting calculus of nomic probabilities is still just a strengthened probability calculus. The strengthenings are important, but not philosophically revolutionary. The main weight of the theory is borne by the acceptance rule. But it deserves to be emphasized that that rule is not revolutionary either. It is just a tidied-up version of the classical statistical syllogism. What makes the theory work where earlier theories did not is the explicit use of defeasible reasoning. It is the importation of state-of-the-art epistemology that infuses new life into the old ideas on which the theory of nomic probability is based.

The most remarkable aspect of the theory of nomic probability is that theories of direct inference and induction fall out as theorems. We need no new postulates in order to justify these essential kinds of probabilistic reasoning. In an important sense, this constitutes a solution to the problem of induction. Of course, you cannot get something from nothing, so we are still making epistemological assumptions that Hume would have found distasteful, but in an important sense principles of induction have been derived from principles that are epistemologically more basic.

# Bibliography

Anscombe, G. E. M.
1958 *Intention.* Ithaca, Cornell University Press.
Ariely, D.
2000 "Seeing sets: Representation by statistical properties," *Psychological Science* **12**, 157–162.
Ariely, D., and C. A. Burbeck
1995 "Statistical encoding of multiple stimuli: A theory of distributed representation," *Investigative Opthalmology and Visual Science* **36** (Suppl.), 8472 (Abstract).
Armendt, Brad
1986 "A foundation for causal decision theory," *Topoi* **5**, 3–19.
1988 "Conditional reference and causal expected utility," W. Harper and B. Skyrms (eds.), *Causation in Decision, Belief Change, and Statistics* **II**, 3–24, Amsterdam, Kluwer.
Bacchus, Fahiem,
1990 *Representing and Reasoning with Probabilistic Knowledge.* Cambridge, Mass., MIT Press.
Bechara, A.
2001 "Neurobiology of decision-making: Risk and reward," *Seminars in Clinical Neuropsychiatry* **6**, 205–216.
2003 "Risky business: Emotion, decision-making and addiction," *Journal of Gambling Studies* **19**, 23–51.
Blum, A., and M. L. Furst
1995 "Fast planning through planning graph analysis," *Proceedings of the Fourteenth International Joint Conference on AI*, 1636–1642.
1997 "Fast planning through planning graph analysis," *Artificial Intelligence* **90**, 281–300.
Blythe, Jim, and Manuela Veloso
1997 "Analogical replay for efficient conditional planning," *Proceedings of AAAI-97*, 668–673.
Bolker, Ethan
1966 "Functions resembling quotients of measures," *Transactions of the American Mathematical Society* **124**, 293–312.
Boutelier, Craig, Thomas Dean, and Steve Hanks
1999 "Decision theoretic planning: Structural assumptions and computational leverage," *Journal of AI Research* **11**, 1–94.
Braithwaite, R. B.
1953 *Scientific Explanation.* Cambridge, Cambridge University Press.
Brosnan, S. F., and F. B. M. de Waal
2003 "Monkeys reject unequal pay," *Nature* **425**, 297–299.

Brosnan, S. F., H. C. Schiff, and F. B. M. de Waal

2005 "Tolerance for inequity may increase with social closeness in chimpanzees," *Proceedings of the Royal Society of London, Series B.* **272**, 253–258.

Cartwright, Nancy

1979 "Causal laws and effective strategies," *Nous* **13**, 419–438.

Catania, A. C.

1963 "Concurrent performances: A baseline for the study of reinforcement magnitude," *Journal of the Experimental Analysis of Behavior* **6**, 299–300.

Cheng, P. W., and K. J. Holyoak

1985 "Pragmatic reasoning schemas," *Cognitive Psychology* **17**, 391–406.

Cherniak, Christopher

1986 *Minimal Rationality.* Cambridge, Mass, MIT Press.

Chomsky, Noam

1957 *Syntactic Structures.* The Hague, Mouton and Co.

Chong, S. C., and A. Treisman

2003 "Representation of statistical properties," *Vision Research* **43**, 393–404.

Church, Alonzo

1940 "On the concept of a random sequence," *Bulletin of the American Mathematical Society* **46**, 130–135.

Danto, Arthur

1963 "What can we do?" *The Journal of Philosophy* **60**, 435–445.

Davidson, Donald

1963 "Actions, reasons, and causes," *The Journal of Philosophy* **60**, 685–700.

de Finetti, B.

1937 "La prevision; ses lois logiques, ses sources subjectives," *Annals de l'Institut Henri Poincare,* vol. 7, 1–68. (Reprinted in 1964 in English translation as "Foresight: Its logical laws, its subjective sources," in *Studies in Subjective Probability,* H. E. Kyburg, Jr., and H. E. Smokler (eds.), New York, John Wiley.)

1974 *Theory of Probability,* vol. 1. New York, John Wiley.

Domasio, Antonio

1994 *Descartes' Error: Emotion, Reason, and the Human Brain.* New York, Putnam's.

Draper, Denise, Steve Hanks, and Daniel Weld

1994 "Probabilistic planning with information gathering and contingent execution," *Proceedings of AIPS94,* 31-36.

Eells, Ellory

1981 "Causality, utility, and decision," *Synthese* **48**, 295–329.

1991 *Probabilistic Causality.* Cambridge, Cambridge University Press.

Fetzer, James

1971 "Dispositional probabilities," *Boston Studies in the Philosophy of Science* **8**, 473–482. Dordrecht, Reidel.

1977 "Reichenbach, references classes, and single case 'probabilities'," *Synthese* **34**, 185–217.

1981 *Scientific Knowledge.* Volume 69 of *Boston Studies in the Philosophy of Science.* Dordrecht, Reidel.

Fodor, Jerry
1975 *The Language of Thought*. New York, Crowell.
Gallistel, C. R., Rochel Gelman, and Sara Cordes
2000 "The cultural and evolutionary history of the real numbers," S. Levinson and P. Jaisson (eds.), *Culture and Evolution*. Cambridge, Mass, MIT Press.
Gelfond, Michael, and Vladimir Lifschitz
1993 "Representing action and change by logic programs," *Journal of Logic Programming* **17**, 301–322.
Gibbard, Alan, and William Harper
1978 "Counterfactuals and two kinds of expected value," C. A. Hooker, J. J. Leach, and E. F. McClennen (eds.), *Foundations and Applications of Decision Theory*. Reidel, Dordrecht, 125–162.
Giere, Ronald
1973 "Objective single case probabilities and the foundations of statistics," Patrick Suppes, Leon Henkin, Athanase Joja, and GR. C. Moisil (eds.), *Logic, Methodology, and Philosophy of Science* **IV**. Amsterdam, North Holland, 67–484.
1973a "Review of Mellor's *The Matter of Chance*," *Ratio* **15**, 149–155.
1976 "A Laplacian formal semantics for single-case propensities," *Journal of Philosophical Logic* **5**, 321–353.
Glymour, Clark
1988 "Psychological and normative theories of causal power and the probabilities of causes," G. F. Cooper and S. Moral (eds.), *Uncertainty in Artificial Intelligence*, San Francisco, Morgan Kaufmann,166–172.
Glymour, C. and G. Cooper
1998 *Computation, Causation, and Discovery*. Cambridge, Mass, MIT Press.
Goldman, Alvin
1970 *A Theory of Human Action*. Princeton, Princeton University Press.
Goodman, Nelson
1955 *Fact, Fiction, and Forecast*. Cambridge, Mass, Harvard University Press.
Haddawy, Peter, and Steve Hanks
1990 "Issues in decision-theoretic planning: Symbolic goals and numeric utilities," *Proceedings of the DARPA Workshop on Innovative Approaches to Planning, Scheduling, and Control*, 48–58.
Halpern, J. Y.
1990 "An analysis of first-order logics of probability," *Artificial Intelligence* **46**, 311-350.
Hanks, Steve, and Drew McDermott
1986 "Default reasoning, nonmonotonic logics, and the frame problem," *Proceedings of AAAI-86*.
1987 "Nonmonotonic logic and temporal projection," *Artificial Intelligence* **33**, 379–412.
Harman, Gilbert
1973 *Thought*. Princeton, Princeton University Press.
Harper, D. G. C.

1982 "Competitive foraging in mallards: Ideal free ducks," *Animal Behavior* **30**, 575–584.

Hitchcock, Christopher

1996 "Causal decision theory and decision-theoretic causation," *Nous* **30**, 508–526.

Howson, Colin and Peter Urbach

1989 *Scientific Reasoning: The Bayesian Approach*. LasSalle, Ill., Open Court.

Jackson, Frank, and John Pargetter

1973 "Indefinite probability statements," *Synthese 26*, 205–215.

Jeffrey, Richard

1965 *The Logic of Decision*. New York, McGraw-Hill.

1970 "Dracula meets Wolfman: Acceptance vs. partial belief," Marshall Swain (ed.), *Induction, Acceptance, and Rational Belief*. Dordrecht, Reidel, 157–185.

1983 *The Logic of Decision, 2nd edition*. Chicago, University of Chicago Press.

Joyce, James

1998 *The Foundations of Causal Decision Theory*. Cambridge, Cambridge University Press.

Kaplan, Mark

1996 *Decision Theory as Philosophy*. Cambridge, Cambridge University Press.

Kautz, H. A., and B. Selman

1996 "Pushing the envelope: Planning, propositional logic, and stochastic search," *Proceedings of AAAI-96*, 1194–1201.

1998 "Blackbox: A new approach to the application of theorem proving to problem solving," *AIPS98 Workshop on Planning as Combinatorial Search*, 58–60.

Keller, J. V., and L. R. Gollub

1977 "Duration and rate of reinforcement as determinants of concurrent responding," *Journal of the Experimental Analysis of Behavior* **28**, 145–153.

Kemeny, John G.

1955 "Fair bets and inductive probabilities," *Journal of Symbolic Logic* **20**, 263–273.

Kiiveri, H., and T. Speed

1982 "Structural analysis of multivariate data: A review," S. Leinhardt (ed.), *Sociological Methodology*. San Francisco, Jossey-Bass.

Kneale, William

1949 *Probability and Induction*. Oxford, Oxford University Press.

Kolmogorov, A. N.

1933 *Grundbegriffe der Wahrscheinlichkeitsrechnung*. Berlin.

Koob, G. F., and N. E. Goeders

1989 "Neuroanatomical substrates of drug self-administration," J. M. Liebman and S. J. Cooper (eds.), *Neuropharacological Basis of Reward*. New York, Oxford University Press, 214–263.

Kushmerick, N., Hanks, S., and Weld, D.

1995 "An algorithm for probabilistic planning," *Artificial Intelligence* **76**, 239–286.

Kyburg, Henry, Jr.

1961 *Probability and the Logic of Rational Belief.* Middletown, Wesleyan University Press.
1970 "Conjunctivitis," Marshall Swain (ed.), *Induction, Acceptance, and Rational Belief.* Dordrecht, D. Reidel.
1974 *The Logical Foundations of Statistical Inference.* Dordrecht, Reidel.

Landauer, Thomas K.
1986 "How much do people remember? Some estimates of the quantity of learned information in long-term memory," *Cognitive Science* **10**, 477–494.

Lane, Richard
2000 "Neural correlates of conscious emotional experience," R. D. Lane and L. Nadel (eds.), *Neuroscience of Emotion.* New York, Oxford University Press, 345–370.

Lehman, R. Sherman
1955 "On confirmation and rational betting," *Journal of Symbolic Logic* **20**, 251–262.

Leon, M. I., and C. R. Gallistel
1998 "Self-stimulating rates combine subjective reward magnitude and subjective reward rate multiplicatively," *Journal of Experimental Psychology: Animal Behavior Processes* **24**, 265–277.

Levi, Isaac
1977 "Direct inference," *Journal of Philosophy* **74**, 5–29.
1980 *The Enterprise of Knowledge.* Cambridge, Mass., MIT Press.
1981 "Direct inference and confirmational conditionalization," *Philosophy of Science* **48**, 532–552.

Lewis, David
1976 "The paradoxes of time travel," *American Philosophical Quarterly* **13**, 145–52.
1979 . "Counterfactual dependence and time's arrow," *Nous* **13**, 455–476.
1980 "A subjectivist's guide to objective chance," R. Jeffrey (ed.), *Studies in Inductive Logic and Probability,* vol. 2. Berkeley, University of California Press, 263–294.
1981 "Causal decision theory," *Australasian Journal of Philosophy* **59**, 5–30.
1994 "Chance and credence: Humean supervenience debugged," *Mind* **103**, 473–490.

Lifschitz, Vladimir
1987 "Formal theories of action," M. L. Ginsberg (ed.), *Readings in Nonmonotonic Reasoning.* Los Altos, Calif., Morgan-Kaufmann.

Luce, R. Duncan, and Howard Raiffa
1957 *Games and Decisions.* New York, Dover.

Martin-Löf, P.
1966 "The definition of random sequences," *Information and Control* **9**, 602–619.
1969 "Literature on von Mises' *Kollektivs Revisited,*" *Theoria* **35**, 12–37.

McInerney, Peter
2006 "Pollock on rational choice and trying," *Philosophical Studies,* **129**, 253–261.

Meek, Christopher, and Clark Glymour
1994 "Conditioning and intervening," *British Journal for the Philosophy of Science* **45**, 1001–1021.
Mellor, D. H.
1969 "Chance," *Proceedings of the Aristotelian Society*, Supplementary Volume, p. 26.
1971 *The Matter of Chance*. Cambridge. Cambridge University Press.
Millgram, Elijah
1997 *Practical Induction*. Cambridge, Mass., Harvard University Press.
Ngo, L., P. Haddawy, and H. Nguyen
1998 "A modular structured approach to conditional decision-theoretic planning," *AIPS'98*, 111–118.
Nozick, Robert
1969 "Newcomb's problem and two principles of choice," N. Rescher (ed.), *Essays in Honor of Carl Hempel*. Dordrecht, Reidel, 107–133.
1974 *Anarchy, State, and Utopia*. New York, Basic Books.
Onder, Nilufer, and Martha Pollack
1997 "Contingency selection in plan generation," *ECP97*, 364–376.
1999 "Conditional, probabilistic planning: A unifying algorithm and effective search control mechanisms," *AAAI 99*, 577–584.
Onder, Nilufer, Martha Pollack, and John Horty
1998 "A unifying algorithm for conditional, probabilistic planning," *AIPS1998 Workshop on Integrating Planning, Scheduling, and Execution in Dynamic and Uncertain Environments*.
Pearl, Judea
2000 *Causality*. Cambridge, Cambridge University Press.
Pednault, Edwin P.
1987 *Toward a Mathematical Theory of Plan Synthesis*. PhD dissertation, Stanford University.
Penberthy, J. Scott, and Daniel Weld
1992 "UCPOP: A sound, complete, partial order planner for ADL," *Proceedings 3rd International Conference on Principles of Knowledge Representation and Reasoning*, 103-114.
Peot, M. A., and D. Smith
1992 "Conditional nonlinear planning," *Proceedings of the First International Conference on Artificial Intelligence Planning Systems*, 189-197.
Pollock, John
1983 "A theory of direct inference," *Theory and Decision* **15**, 29–96.
1983a "Epistemology and probability," *Synthese* **55**, 231–252.
1984 "Foundations for direct inference," *Theory and Decision* **17**, 221–256.
1984a "A solution to the problem of induction," *Nous* **18**, 423–462.
1986 *Contemporary Theories of Knowledge*. Totowa, N.J., Rowman and Littlefield.
1987 "Probability and proportions," H. Berghel and G. Eberlein (eds.), *Theory and Decision: Essays in Honor of Werner Leinfellner*. Amsterdam, Reidel.

1989 *How to Build a Person.* Cambridge, Mass., Bradford/MIT Press.
1990 *Nomic Probability and the Foundations of Induction.* New York, Oxford University Press.
1992 "New foundations for practical reasoning," *Minds and Machines* **2**, 113–144.
1992a "The theory of nomic probability," *Synthese* **90**, 263–300.
1995 *Cognitive Carpentry.* Cambridge, Mass., Bradford/MIT Press.
1998 "The logical foundations of goal-regression planning in autonomous agents," *Artificial Intelligence* **106**, 267–335.
1998a "Perceiving and reasoning about a changing world," *Computational Intelligence* **14**, 498–562.
2001 "Evaluative cognition," *Nous*, **35**, 325–364.
2002 "Causal probability," *Synthese* **132**, 143–185.
2002a "Defeasible reasoning with variable degrees of justification," *Artificial Intelligence* **133**, 233–282.
2003 "Rational choice and action omnipotence," *Philosophical Review* **111**, 1–23.
2007 "Irrationality and cognition," Quentin Smith (ed.), *Epistemology: New Philosophical Essays.* New York, Oxford University Press.
2007a "Defeasible reasoning," Jonathan Adler and Lance Rips (eds.), *Reasoning: Studies of Human Inference and its Foundations.* Cambridge, Cambridge University Press.
2007b "Against optimality: Logical foundations for decision-theoretic planning in autonomous agents," *Computational Intelligence* **22**, 1-25.

Pollock, John, and Joseph Cruz
1999 *Contemporary Theories of Knowledge, 2nd edition.* Lanham, Md., Rowman and Littlefield.

Popper, Karl
1956 "The propensity interpretation of probability," *British Journal for the Philosophy of Science* **10**, 25–42.
1957 "The propensity interpretation of the calculus of probability, and the quantum theory," S. Körner (ed.), *Observation and Interpretation.* New York, Academic Press, 65–70.
1959 *The Logic of Scientific Discovery.* New York, Basic Books.
1960 "The propensity interpretation of probability," *British Journal for the Philosophy of Science* **11**, 25–42.

Price, Huw
1991 "Agency and probabilistic causality," *British Journal for the Philosophy of Science* **42**, 157–176.

Pryor, Louise, and Gregg Collins
1996 "Planning for contingencies: a decision-based approach," *Journal of Artificial Intelligence Research* **4**, 287-339.

Pylyshyn, Zenon
1999 "Is vision continuous with cognition? The case for cognitive

impenetrability of visual perception," *Behavioral and Brain Sciences* **22**, 341–423.

Ramsey, Frank

1926 "Truth and probability," R. B. Braithwaite (ed.), *The Foundations of Mathematics*. Paterson, N.J., Littlefield, Adams.

Reichenbach, Hans

1949 *A Theory of Probability*. Berkeley, University of California Press. (Original German edition 1935)

Russell, Bertrand

1948 *Human Knowledge: Its Scope and Limits*. New York, Simon and Schuster.

Salmon, Wesley

1977 "Objectively homogeneous references classes," *Synthese* **36**, 399–414.

Russell, Stuart, and Peter Norvig

1995 *Artificial Intelligence: A Modern Approach*. Upper Saddle River, N.J., Prentice Hall.

Savage, Leonard

1954 *The Foundations of Statistics*. New York, Dover.

Schnorr, C. P.

1971 "A unified approach to the definition of random sequences," *Mathematical Systems Theory* **5**, 246–258.

Schwayder, David

1965 *The Stratification of Behavior*. New York, Humanities Press.

Shanahan, Murray

1990 "Representing continuous changes in the event calculus," *ECAI-90*, 598–603.

1995 "A circumscriptive calculus of events," *Artificial Intelligence*, **77**, 249–284.

1996 "Robotics and the common sense informatic situation," *Proceedings of the 12th European Conference on Artificial Intelligence*. New York, John Wiley.

1997 *Solving the Frame Problem*. Cambridge, Mass., MIT Press.

Shimony, Abner

1955 "Coherence and the axioms of confirmation," *Journal of Symbolic Logic* **20**, 1–28.

Shoham, Yoav

1986 *Time and Causation from the Standpoint of Artificial Intelligence*, Computer Science Research Report No. 507. New Haven, Yale University.

1987 *Reasoning about Change*, MIT Press.

Simon, Herbert

1955 "A behavioral model of rational choice," *Quarterly Journal of Economics* **59**, 99–118.

Sklar, Lawrence

1970 "Is propensity a dispositional concept?" *Journal of Philosophy 67*, 355–366.

1973 "Unfair to frequencies," *Journal of Philosophy* **70**, 41–52.

1974 "Review of Mellor's *The Matter of Chance*," *Journal of Philosophy* **71**, 418–423.

Skyrms, Brian

1980 *Causal Necessity*. New Haven, Yale University Press.

1982 "Causal decision theory," *Journal of Philosophy* **79**, 695–711.

1984 *Pragmatics and Empiricism*. New Haven, Yale University Press.

Sobel, Jordan Howard

1978 "Probability, chance, and choice: A theory of rational agency," unpublished paper presented at a workshop on Pragmatism and Conditionals at the University of Western Ontario, May, 1978.

1994 *Taking Chances: Essays on Rational Choice*. New York, Cambridge University Press.

Spirites, P., Clark Glymour, and Richard Scheines

1993 *Causation, Prediction, and Search*. New York, Springer-Verlag.

Stalnaker, Robert

1978 "Letter to David Lewis," reprinted in W. Harper, R. Stalnaker, G. Pearce, (eds.), *Ifs*, Dordrecht, Reidel.

Stalker, Douglas

1994 *Grue: The New Riddle of Induction*. Chicago, Open Court.

Suppes, Patrick

1970 *A Probabilistic Theory of Causality*. Amsterdam, North-Holland.

1973 "New foundations for objective probability: Axioms for propensities," Patrick Suppes, Leon Henkin, Athanase Joja, and GR. C. Moisil (eds.), *Logic, Methodology, and Philosophy of Science* **IV**. Amsterdam, North Holland, 515–529.

1988 "Probabilistic causality in space and time," B. Skyrms and W. L. Harper (eds.), *Causation, Chance, and Credence*. Dordrecht, Kluwer.

Taylor, G. J., R. R. Bagby, and J. D. A. Parker

1991 "The alexithymia construct: A potential paradigm for psychosomatic medicine," *Psycosomatics* **32**, 153–164.

Tranel, D.

1997 "Emotional processing and the human amygdala," *Trends in Cognitive Sciences* **1**, 99–113.

Venn, John

1888 *The Logic of Chance, 3rd edition*. London, Macmillan.

von Mises, Richard

1957 *Probability, Statistics, and Truth*. New York, Macmillan. (Original German edition 1928)

von Neumann, J., and O. Morgenstern

1944 *Theory of Games and Economic Behavior*. New York, Wiley.

Warren, D.

1976 "Generating conditional plans and programs," *Proceedings of the Summer Conference and AI and Simulation of Behavior*.

Wason, P.
1966 "Reasoning," B. Foss (ed.), *New Horizons in Psychology*. Harmondsworth, England, Penguin.

Weld, Daniel
1999 "Recent advances in AI planning," *AI Magazine* **20**, 55–68.

Williamson, Mike, and Steve Hanks
1994 "Beyond symbolic goals: A step toward utility-directed planning," *AAAI Spring Symposium on Decision-Theoretic Planning*, 272–278.

# Index